146
Advances in Polymer Science

Springer-Verlag Berlin Heidelberg GmbH

Polymer Synthesis
Polymer-Polymer Complexation

With contributions by
S. Inoue, S. Jacob, M. Jiang,
J. P. Kennedy, M. Li, H. Sugimoto,
M. Xiang, H. Zhou

 Springer

This series presents critical reviews of the present and future trends in polymer and biopolymer science including chemistry, physical chemistry, physics and materials science. It is addressed to all scientists at universities and in industry who wish to keep abreast of advances in the topics covered.

As a rule, contributions are specially commissioned. The editors and publishers will, however, always be pleased to receive suggestions and supplementary information. Papers are accepted for „Advances in Polymer Science" in English.

In references Advances in Polymer Science is abbreviated Adv. Polym. Sci. and is cited as a journal.

Springer WWW home page: http://www.springer.de

ISSN 0065-3195
ISBN 978-3-662-15617-9   ISBN 978-3-540-49424-9 (eBook)
DOI 10.1007/978-3-540-49424-9

Library of Congress Catalog Card Number 61642

© Springer-Verlag Berlin Heidelberg 1999
Originally published by Springer-Verlag Berlin Heidelberg New York in 1999
Softcover reprint of the hardcover 1st edition 1999

Typesetting: Data conversion by MEDIO, Berlin
Cover: E. Kirchner, Heidelberg
SPIN: 10691439   02/3020 - 5 4 3 2 1 0 - Printed on acid-free paper

Editorial Board

Contents

Synthesis, Characterization and Properties of Octa-Arm Polyisobutylene-Based Star Polymers

Sunny Jacob, Joseph P. Kennedy

The Maurice Morton Institute of Polymer Science, The University of Akron, Akron, Ohio 44325–3909, USA
e-mail: moyers@polymer.uakron.edu

This review concerns the synthesis, characterization, and select properties of a series of novel eight-arm polyisobutylene (PIB)-based stars. An octafunctional initiator containing 2-methoxyisopropyl groups 1 was synthesized by quantitative functionalization of 5,11, 17,23,29,35,41,47-octaacetyl-49,50,51,52,53,54,55,56-octamethoxy-calix[8]arene and used to prepare the stars. Stars with PIB arms, terminally functionalized PIB arms, poly(styrene-*b*-isobutylene) (PSt-*b*-PIB) arms, and poly(*p*-chlorostyrene-*b*-isobutylene) (P*p*ClSt-*b*-PIB) arms were synthesized. Polymerizations of isobutylene (IB) to prepare well-defined PIB stars were effected by the use of octa-functional initiator with BCl_3-$TiCl_4$ coinitiators in two stages. Stage I was carried out in the presence of a fraction of the required amount of IB in CH_3Cl and polymerization was induced by the addition of BCl_3. Stage II was induced by the addition of $TiCl_4$ and the balance of IB, and the polymerization was carried to completion. The relative concentrations of BCl_3 and $TiCl_4$ were critical for the synthesis of well-defined stars. By the use of well-chosen BCl_3-$TiCl_4$ concentrations, added in sequence in two stages, narrow dispersity stars, $\overline{M}_w / \overline{M}_n = 1.11$, were prepared. Gel permeation chromatography (GPC), with on-line RI, UV, and laser light scattering (LLS) detectors, was used to determine the molecular weights and compositions of linear and octa-arm star polymers. The number of arms of the stars was determined by ^1H NMR spectroscopy and by selective core destruction. Both studies indicated that the number of arms was close to the theoretical value of eight. The mechanism of polymerization was studied by the use of a model monofunctional analogue of the octa-functional initiator, 2-(*p*-methoxyphenyl)-2-methoxypropane. Allyl-end functionalized octa-arm PIB stars were prepared by quenching the living PIB$^\oplus$ stars by allyltrimethylsilane (ATMS). Star thermoplastic elastomers containing PSt-*b*-PIB and P*p*ClSt-*b*-PIB arms were synthesized by sequential addition of styrene and *p*-chlorostyrene, respectively, to the living PIB$^\oplus$ stars. These star blocks contained 10–15% PIB and/or diblocks and showed an excellent combination of thermoplastic elastomeric properties, thermal properties, and melt flow properties.

1
Introduction

A star polymer is a polymer consisting of several linear chains (arms) connected together at one end to a common point (core) [1–6]. In regular or symmetric stars, the arms (homopolymers or block copolymers) have the same chemical composition and molecular weight, and in hetero-arm stars or mikto-arm stars, the branches have different molecular weights and/or chemical compositions [6]. The synthesis of well-defined stars with a known number of arms, arm molecular weights, and narrow dispersities is a challenge to the macromolecular engineer. The application of living polymerization methods has greatly helped the synthesis of well-defined stars.

Living polymerization methods offer the best routes to the synthesis of well-defined stars [3–5,7]. Linking the living chains to a core or growing living chains from a core are two ways for the synthesis of stars. The core may be prepared separately or formed in situ during linking the arms.

The synthesis of well-defined stars was first accomplished by living anionic polymerization [8]. Later living cationic [7], group transfer [9], living/controlled free radical [10], and other methods [11] have also been used. Stars can be prepared by core-first, arm-first, and arm-core linking methods. In the core-first method, a multifunctional initiator (the residue of which becomes the core) with a certain number of initiating sites is prepared, and this core is then used to initiate the living polymerization of monomers. Under optimum conditions the initiating sites induce the polymerization and lead to stars with uniform arms of known composition. This method has been used to prepare star polymers by anionic [6, 12–17], cationic [7, 18–27], GTP [9, 28], radical [10, 29], and condensation [30, 31] polymerizations. One important advantage of the core-first method is that the reactivity of the chain ends is not lost during star formation, which renders the preparation of star blocks and functionalized-stars easy. The number of arms is limited by the number of initiating sites in the core.

Stars with high arm numbers are commonly prepared by the arm-first method. This procedure involves the synthesis of living precursor arms which are then used to initiate the polymerization of a small amount of a difunctional monomer, i.e., for linking. The difunctional monomer produces a crosslinked microgel (nodule), the core for the arms. The number of arms is a complex function of reaction variables. The arm-first method has been widely used in anionic [3–6, 32–34], cationic [35–40], and group transfer polymerizations [41] to prepare star polymers having varying arm numbers and compositions.

In the arm-core linking method, living chains or functionalized chains are prepared by living polymerization of a monomer and are subsequently reacted with multifunctional linking reagents containing a known number of reactive sites to form the star. The maximum number of arms that can be obtained is predetermined by the nature of the multifunctional reagent. Since most linking reactions involving multifunctional linking agents are diffusion controlled, long reaction times and/or forced conditions are required for completion of the synthesis. This method has been used to prepare well-defined star homopolymers, star block co-polymers, and hetero-arm stars by anionic [3, 6–8, 44–52], GTP [9, 51], cationic [52–55], and by combination of anionic and condensation [56, 57] methods.

Although the core-first method is the simplest, success depends on initiator preparation and quantitative initiation under living conditions. This method is of limited use in anionic polymerization because of the generally poor solubility of multifunctional initiators in hydrocarbon solvents [12]. Solubilities of multi-functional initiators are less of an issue in cationic polymerizations, and tri- and tetrafunctional initiators have been used to prepare well-defined three- and four-arm star polymers by this method [7] Except for two reports on the synthe-sis of hexa-arm polystyrene [27] and hexa-arm polyoxazoline [26], there is a dearth of information in regard to well-defined multifunctional initiators for the preparation of higher functionality stars.

This review concerns the synthesis and characterization of octa-arm polyisobutylene (PIB) stars, allyl-terminated octa-arm PIB stars, and octa-arm star blocks by using a novel octafunctional calix[8]arene-based initiator 1. Scheme 1 shows the structure of 1 and the target architectures. The syntheses were carried out under living carbocationic polymerization conditions.

Calix[n]arenes (n=4, 6, 8) are cyclic condensation products of a p-substituted phenol and formaldehyde [58]. Gutsche and co-workers [59, 60] have developed procedures for the synthesis of calixarenes and calixarene derivatives.

2
Experimental

2.1
Materials

The synthesis and characterization of 1 have been described [61, 62]. Isobuty-lene [IB (CP grade)] and methyl chloride were obtained from Matheson. The gases were dried by passing through in-line columns packed with BaO/Dri-erite/molecular sieves/$CaCl_2$. Methylcyclohexane (MeCH) (Aldrich) was reflux-ed overnight in the presence of CaH_2 and distilled before use. Styrene, St (Aldrich), and p-chlorostyrene, pClSt (Lancaster) were purified by passing these monomers through inhibitor remover columns before use. Other chemicals (Aldrich) were used as received.

2.2
Polymerizations

2.2.1
Preparation of Octa-Arm Polyisobutylenes

Polymerizations were carried out either in 75-ml culture tubes or in 250-ml stirred reactors in two stages according to a published procedure [61, 62]. In

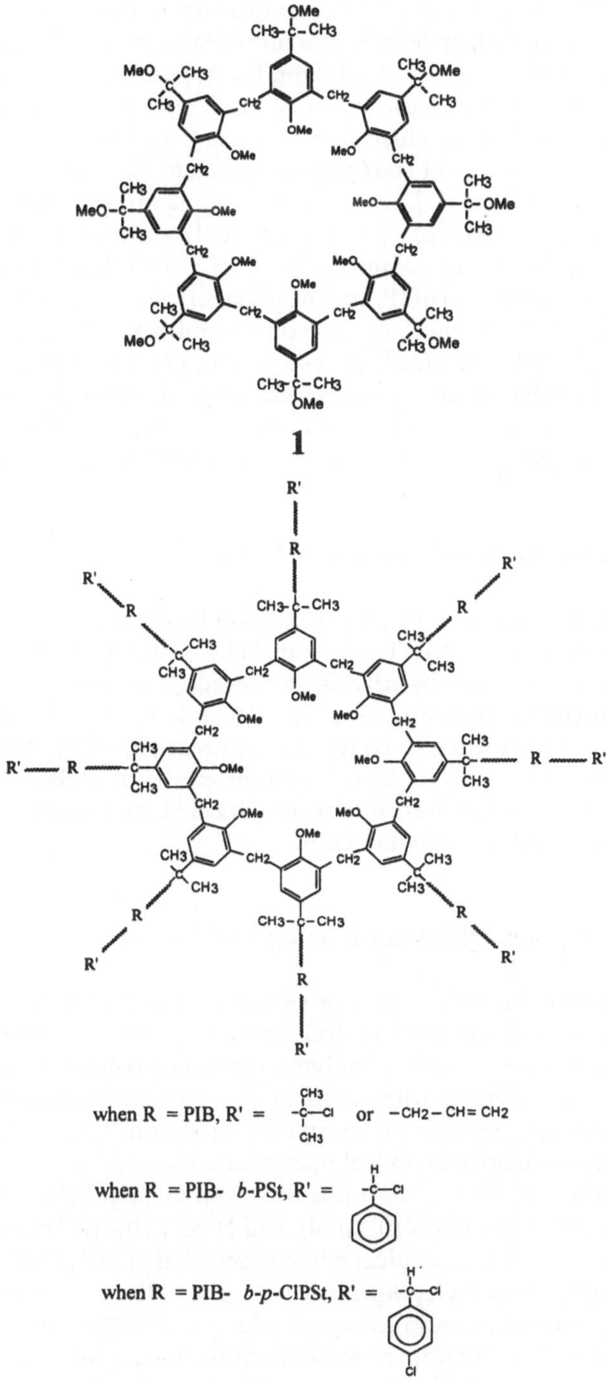

1

when R = PIB, R' =

$$-\overset{CH_3}{\underset{CH_3}{\overset{|}{C}}}-Cl \quad \text{or} \quad -CH_2-CH=CH_2$$

when R = PIB- b-PSt, R' =

$$-\overset{H}{\underset{}{\overset{|}{C}}}-Cl$$

when R = PIB- b-p-ClPSt, R' =

$$-\overset{H}{\underset{}{\overset{|}{C}}}-Cl$$

Scheme 1

stage I, the addition sequence of the reactants was: initiator 1, CH_3Cl, IB (~25% of the required amount), dimethylacetamide (DMA), di-*tert*-butyl-pyridine (Dt-BP), and BCl_3. Stock solutions of initiator, DMA, DtBP, and BCl_3 were prepared in CH_3Cl and $TiCl_4$ in hexanes. After polymerizing for 60 min, hexanes, $TiCl_4$, and the balance of IB were added. The ultimate CH_3Cl:hexanes ratio was 40:60 (v/v). High molecular weight stars were prepared by the incremental monomer addition (IMA) technique [7]. A representative polymerization was carried out as follows. Initiator 1, 4.81×10^{-2} g (3.15×10^{-5} mol), was placed in a 75-ml culture tube. In stage I, the charge contained 10 ml of CH_3Cl, DMA (5.01×10^{-4} mol), DtBP (3.19×10^{-4} mol), 1 ml of IB, and the polymerization was induced by the addition of BCl_3 (7.54×10^{-4} mol). After 60 min, 15 ml of hexanes, $TiCl_4$ (2.30×10^{-3} mol) and 3.4 ml of IB were added. After 45 min the polymerization was quenched by pre-chilled methanol. The solvents were evaporated, the polymer was redissolved in hexanes, the hexanes layer was washed with 5% HCl, water, and methanol, and the polymer was dried in vacuum. Conversion: 3.2 g (100%).

2.2.2
Preparation of Allyl-Functionalized Octa-Arm PIB Stars

The overall methodology of IB polymerization by the two-stage procedure has been followed (see Sect. 2.2.1). The terminal allyl-functionalization of the living $PIB^{\oplus}$ arms was carried out by adapting a procedure developed in these laboratories [63]. Allyltrimethylsilane (ATMS) (~100-fold excess relative to the $PIB^{\oplus}$ chain end) was added when IB conversion has reached ~95%. After 60 min the polymerization mixture was poured into an excess of chilled methanol, the product separated, redissolved in hexanes, washed with water and methanol, and finally reprecipitated into methanol.

2.2.3
Preparation of Octa-Arm Poly(styrene-b-isobutylene) Stars

The synthesis of living $PIB^{\oplus}$ stars was similar to that used in Sect. 2.2.1. IB was polymerized to predetermined molecular weights and after reaching at least 95% conversion, St was added and polymerized for a predetermined time to prepare star-block copolymers with varying polystyrene (PSt) compositions. Samples (0.5–1 ml) were withdrawn to follow conversions and molecular weight build-up. A representative block polymerization was carried out as follows: Initiator 1, (0.0491 g, 3.19×10^{-5} mol) was dissolved in CH_3Cl (10 ml), then, in sequence, IB (1 ml), DMA (5.0×10^{-4} mol), and DtBP (3.0×10^{-4} mol) were added, and the polymerization was induced by the addition of BCl_3 (7.65×10^{-4} mol) at −80 °C. After 60 min methylcyclohexane (15 ml) and IB (3.2 ml) were added and the polymerization was continued by the addition of $TiCl_4$ (2.04×10^{-3} mol). After 80 min and after 110 min of total reaction time, a mixture of methylcyclohexane (9 ml), CH_3Cl (6 ml) and IB (4.2 ml) was added followed by $TiCl_4$ (1.22×10^{-3} mol). After 140 min a mixture of methylcyclohexane (21 ml) and

CH$_3$Cl (14 ml), and styrene (6.34 ml) was added and a further amount of TiCl$_4$ (1.22×10^{-3} mol) was introduced. After 155 min the polymerization was quenched by the use of pre-chilled methanol. The product was washed with 5% HCl, water, and methanol, and the volatiles removed by evaporation. It was purified by dissolving in THF and reprecipitating by adding methanol and dried in vacuum at room temperature. Stars with different PSt content were prepared by increasing the blocking time and/or by decreasing the molecular weight of the PIB segments.

2.2.4
Preparation of Octa-Arm Poly(p-chlorostyrene-b-isobutylene) Stars

The synthesis was carried out as described in Sect. 2.2.3 except that *p*-ClSt was used instead of St. Initiator 1 (0.0491 g, 3.19×10^{-5} mol) was dissolved in CH$_3$Cl (10 ml), then, in sequence, IB (1 ml), DMA (5.0×10^{-4} mol), and DtBP (3.0×10^{-4} mol) were added, and the polymerization was induced by the addition of BCl$_3$ (7.65×10^{-4} mol) at −80 °C. After 60 min methylcyclohexane (15 ml) and IB (3.2 ml) were added and the polymerization was continued by the addition of TiCl$_4$ (2.04×10^{-3} mol). After 80 and 110 min of total reaction time a mixture of methylcyclohexane (9 ml), CH$_3$Cl (6 ml) and IB (4.2 ml) was added followed by TiCl$_4$ (1.22×10^{-3} mol). After 140 min a mixture of methylcyclohexane (36 ml) and CH$_3$Cl (24 ml), and *p*-ClSt (5.0 ml) was added and a further amount of TiCl$_4$ (3.06×10^{-3} mol) was introduced. After 295 min the polymerization was quenched by the use of prechilled methanol. The product was washed with 5% HCl, water, and methanol, and the volatiles removed by evaporation. It was purified by dissolving in THF and reprecipitating by adding methanol and dried in vacuum at room temperature. Stars with different P*p*ClSt content were prepared by increasing the blocking time, by increasing the polarity during blocking and/or by decreasing the molecular weight of the PIB segments.

2.3
Fractionation of PIB Stars

PIB star (3 g) was dissolved in 200 ml of hexanes. Acetone was added drop by drop into the solution until a persistent cloudiness appeared. The mixture was then gently heated (~40 °C) so that a clear solution was obtained. To the hot clear solution more acetone was added until the charge became cloudy. The container was sealed and left undisturbed overnight. The star product separated at the bottom was collected. The procedure was repeated three to four times.

2.4
Solvent Extraction of Star Blocks

To determine the amount of homopolymer contaminants, select samples were solvent extracted. Thus 5 g of the star block was placed in a cellulose thimble and extracted with boiling methylethyl ketone, MEK (a good solvent for PSt and P*p*-ClSt) for 48 h in a Soxhlet apparatus. The percent of extractables was determined by gravimetry.

2.5
Core Destruction of Star Polymers

Star polymer [$\overline{M}_w$(LLS)=1.159×10^5 g/mol, entry 2; Table 3], 0.26 g, was dissolved in 25 ml CCl_4 in a 250-ml two-neck round bottom flask fitted with a condenser and N_2 inlet. A mixture of 14 ml of trifluoroacetic acid and 4 ml of 30% aqueous H_2O_2 was added with stirring. The charge was heated under reflux (75–80 °C); samples were withdrawn at 2-h intervals, quenched by methanol, evaporated to dryness, redissolved in hexanes, precipitated using methanol, and dried in vacuum. GPC analysis of a sample withdrawn after 16 h indicated a single peak (i.e., complete core destruction). $\overline{M}_w$(LLS)=1.496×10^4 g/mol.

Control experiments were carried out under identical conditions by the use of a linear PIB ($\overline{M}_n$~10,000 g/mol, $\overline{M}_w$ / $\overline{M}_n$=1.2) and 1. GPC analysis indicated that the calixarene was destroyed and formed low molecular weight products, whereas the PIB survived the oxidation.

3
Characterization

^{1}H NMR and ^{13}C NMR spectra (~30 and ~50 mg samples, respectively) were recorded by a Varian Gemini-200 spectrometer using standard 5-mm tubes at room temperature. For ^{1}H NMR spectroscopy 64 FIDs were collected and for ^{13}C NMR spectroscopy more than 4000 FIDs were collected.

Molecular weights were determined by GPC (Waters Co.) equipped with a series of five μ-Styragel columns (100, 500, 10^3, 10^4, and 10^5 Å), an RI detector (Waters 410 Differential Refractometer), a UV detector (440 Absorbance Detector), and a laser light scattering (LLS) detector (Wyatt Technology). The columns were calibrated using narrow molecular weight PIB standards. The dn/dc values were obtained by using an Optilab 903 (Wyatt Technology) instrument. Astra software (version 4.00, Wyatt Technology Corporation) was used for data analysis.

Stress-strain measurements were carried out on molded films (~1 mm thickness) at room temperature by the use of microdumbell-shaped samples and an Instron tensile tester (model No. 1130, crosshead speed of 5 cm/min). The samples were premolded between Mylar sheets for 10 min at 162 °C at about 5000 psi, then remolded at 165 °C and 7000 psi for 20 min, and slowly cooled (~1 °C/min) to 50 °C. Select samples were solvent extracted by MEK using a Soxhlet extractor before molding.

A differential scanning calorimeter (DSC), Dupont Instrument, Model DSC2910, was used to determine the glass transition temperatures. Thermogravimetric analyses were carried on a thermogravimetric analyzer (TGA), TA Instruments, Model Hi-Res TGA 2950.

Dynamic melt viscosity studies on the star blocks and a similar triblock were carried out using a Rheometric Mechanical Spectrometer (RMS) (Rheometrics 800). Circular molded samples with ~1.5 mm thickness and 2 cm diameter were subjected to forced sinusoidal oscillations (2% strain) between two parallel plates. The experiment was set in the frequency sweep mode. Data were collected at 180 and 210 °C.

Morphology of select star blocks was investigated by transmission electron microscopy (TEM). Films were cast from toluene and annealed for 2 days at 120 °C. Ultra thin sections (~50 nm) of unstained samples were cut by cryogenic microtome techniques. Samples were viewed by a JOEL (JEM-1200EX II) TEM.

4
Results and Discussion

The synthesis and characterization of well-defined novel octa-arm PIB stars, octa-arm allyl-functionalized PIB stars, and octa-arm star-block copolymers (PSt-b-PIB and PpClSt-b-PIB) wherein the arms emanate from a calix[8]arene core are reviewed. The preparative strategy included precision synthesis of initiator 1 carrying eight initiating sites [61, 62]. The synthesis of octa-arm star PIBs was accomplished by living polymerization of IB using 1 in conjunction with BCl_3 and $TiCl_4$ coinitiators in two stages (see Scheme 2). Allyl end-functionalization of the living $PIB^{\oplus}$ arms was achieved by the use of ATMS as the end quenching agent. PIB-based star blocks were prepared by sequential monomer addition of St or pClSt to the living $PIB^{\oplus}$ arms.

4.1
Synthesis and Characterization of the Octa-Arm PIB Stars

Systematic investigations showed that monomodal octa-arm PIB stars can be obtained from 1 by a two-stage procedure [61, 62]. Thus IB was polymerized in two stages with BCl_3-$TiCl_4$ coinitiators at –80 °C (see Scheme 2). Stage I was carried out in the presence of a fraction of the required amount of IB in CH_3Cl and the polymerization was induced by the addition of BCl_3. During this stage only very low conversions and very low molecular weight products were obtained which indicated that only a few units of IB were added to the initiator. In stage II hexanes, the balance of IB, and $TiCl_4$ were added and the polymerization was carried to completion.

To prove that under these conditions, the IB polymerization is living, a monofunctional analogue of 1, 2-p-methoxyphenyl-2-methoxypropane, was used to study the kinetics by incremental monomer addition technique. Results of this study indicated living polymerization with slow initiation [61, 62].

In stage I it is assumed that 1 produces the initiating cation by the following sequence of transformations (see Scheme 3). The $BCl_3OMe^{\ominus}$ counter-anion is less stable than $BCl_4^{\ominus}$ and rapidly produces the tert-chlorine derivative, which in conjunction with the excess BCl_3 in the charge produces the initiating species,

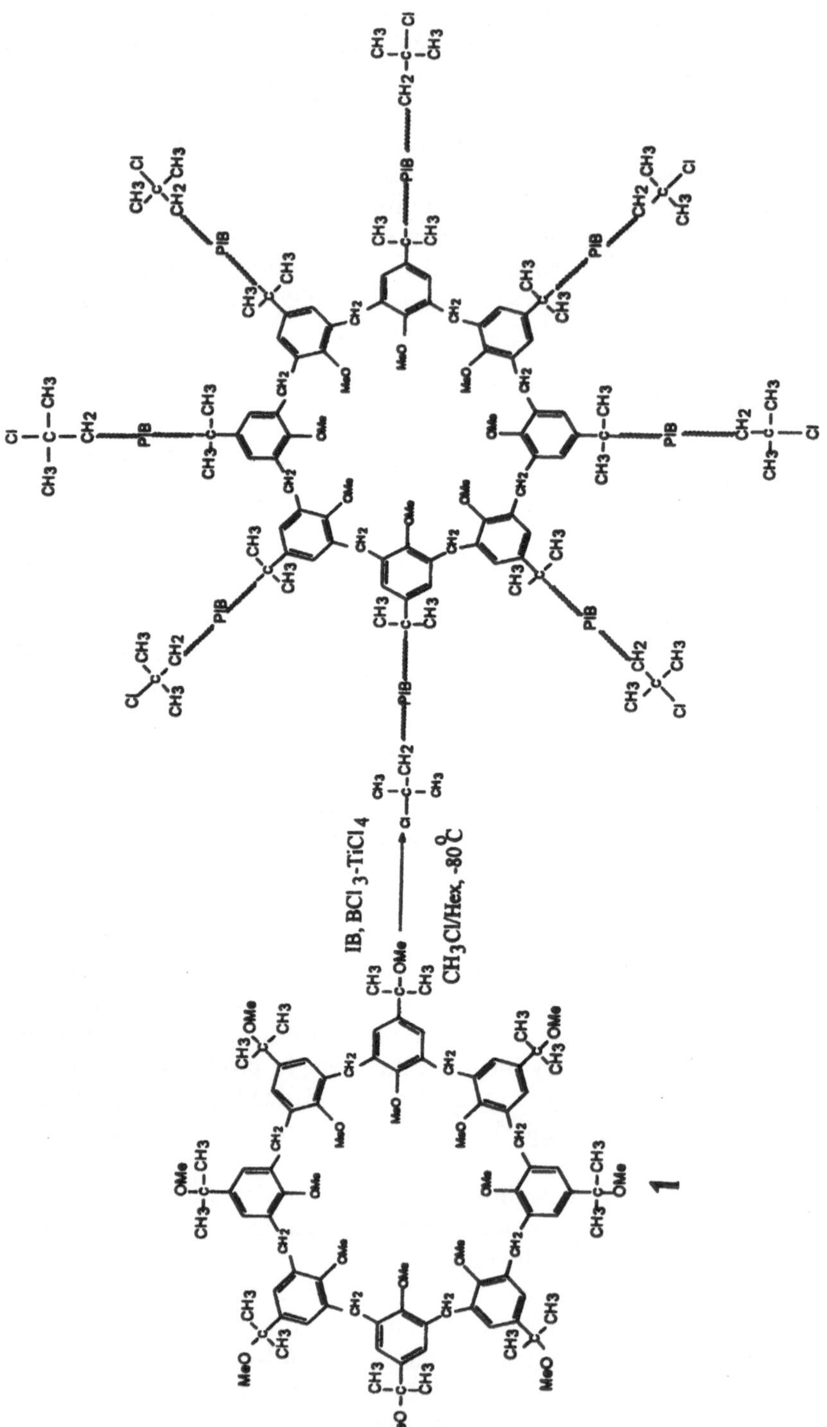

IB, BCl$_3$–TiCl$_4$

CH$_3$Cl/Hex, -80°C

1

Scheme 2

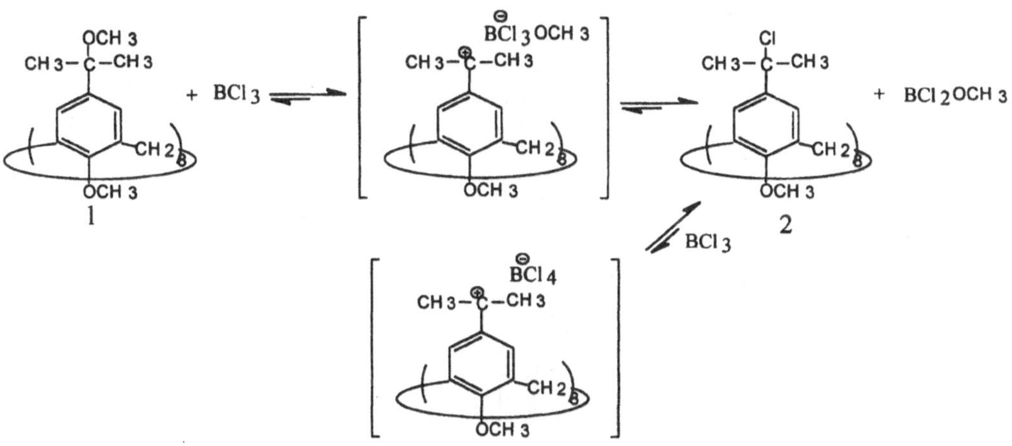

Scheme 3

which initiate the polymerization of IB to form star oligomers. In stage II these living oligomers undergo rapid polymerization in the presence of $TiCl_4$ to the final octa-arm stars.

The use of either BCl_3 or $TiCl_4$ produces unsatisfactory products. BCl_3 is an efficient coinitiator for 1, but since it can initiate IB polymerization only in CH_3Cl, its use is limited to the preparation of low molecular weight stars. The use of $TiCl_4$ alone leads to heterogeneous products. Details of these investigations are published elsewhere [61, 64]. See also Sect. 4.4. In contrast, well-defined narrow dispersity stars, $\overline{M}_w / \overline{M}_n = 1.11$ can be prepared by inducing the polymerization with relatively low concentrations of BCl_3 in Stage I and moderate concentrations of $TiCl_4$ in stage II. Figure 1 shows the GPC (RI) traces of three select samples with (a) $\overline{M}_w(LLS)=67,000$ g/mol, (b) $\overline{M}_w(LLS)= 115,900$ g/mol, and (c) $\overline{M}_w(LLS)=273,000$ g/m, respectively. The minor peaks observed at high elution volumes (~10% by RI peak area) are due to a linear PIB by-product, due to adventitious initiation of IB by the chloroboration product of IB and BCl_3 ($Cl_2B-CH_2-C(CH_3)_2-Cl$) [61, 64].

Pure stars can be readily obtained by removing the linear PIB contaminant by dissolving the mixture in hexanes and precipitating with acetone. Star polymers thus obtained were used for determining the molecular weights by light scattering. Table 1 summarizes select results and Fig. 2 shows the GPC (RI and LLS) traces of a representative sample. Both the LLS (90°) and the RI traces indicate the presence of a narrow dispersity monomodal star. The slight shift of the LLS peak toward higher molecular weights is due to the relatively greater scattering by high molecular weight stars. The dn/dc of the stars was 0.118 cm^3/g. The $\overline{M}_n$, $\overline{M}_w$, and $\overline{M}_w/ \overline{M}_n$, of this star were 1.05×10^5 g/mol, 1.16×10^5 g/mol, and 1.11, respectively (see Table 1). The observed molecular weights of the arms, $\overline{M}_{n,obsd}=13,100$ g/mol (based on the theoretical number of arms), were slightly higher than the theoretical value ($\overline{M}_{n,theor}=11,500$ g/mol

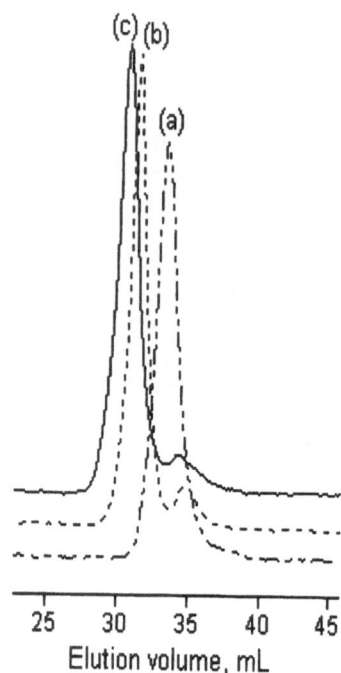

Fig. 1. Gel permeation chromatograms (RI traces) of stars with different molecular weight: *a* $\overline{M}_w$(LLS)=67,000 g/mol; *b* $\overline{M}_w$(LLS)=115,900 g/mol; *c* $\overline{M}_w$(LLS)=273,000 g/mol

Table 1. Molecular characteristics of select octa-arm PIB stars[a]

Initiator, 1	IB addition		Conv.	$\overline{M}_n$, (LLS)[b]	$\overline{M}_w$, (LLS)	$\overline{M}_w / \overline{M}_n$	$\overline{M}_{n,arms}$	
mol×10⁻⁵	ml	time, min	%	g/mol	g/mol		Obsd. (LLS), g/mol	Theor.[c] g/mol
3.18	1	0						
	1.7	60						
		105	~100	62,000	67,000	1.08	7,800	6,800
3.13	1	0						
	3.4	60						
		105	~100	104,800	115,900	1.11	13,100	11,500
3.16	1	0						
	3.5	60						
	4	80						
	4	105	~90	244,000	273,000	1.12	30,500	29,000
		130						

[a] Concentration of initiating sites=1×8, [BCl₃]=3×(1×8), [TiCl₄]=8×(1×8), [DMA]=2×(1×8), [DtBP]=1.03×10⁻² mol/l
[b] Calculated by Astra 4.0
[c] Calculated assuming arm number=8

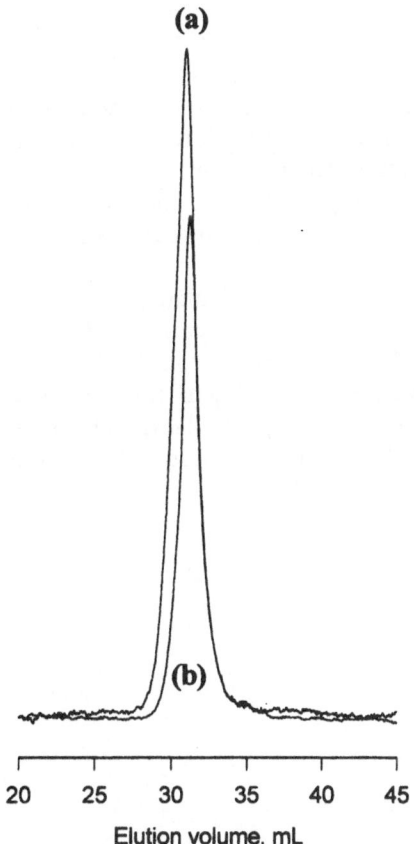

Fig. 2. Gel permeation chromatograms: *a* LLS (90°) trace; *b* RI trace of star polymer (entry 2; Table 1) after fractionation

per arm, $\overline{M}_{n,theor}$ calculated after correction for side-product).The cause of this slight discrepancy may be due to less than 100% initiation efficiency and/or overestimation of the $\overline{M}_n$.

4.2
Determination of Number of Arms in Star Polymers

To verify the expectation that the eight *tert*-methoxy groups in **1** in fact gave rise to eight PIB arms, the stars prepared were analyzed for the number of arms present by two methods. In the first method low molecular weight PIB stars were analyzed by ^{1}H NMR spectroscopy which gives directly the number of arms of stars. In the second method the number of arms was determined indirectly by selective destruction of the aromatic cores followed by GPC analysis.

4.2.1
¹H NMR Spectroscopy

Low-molecular weight stars were prepared to facilitate the interpretation of ¹H NMR spectra. Figure 3 shows the ¹H NMR spectrum of a virgin sample indicating resonances at $\delta=1.95$ and $\delta=1.65$ ppm, characteristic of protons of the terminal -CH$_2$-C(CH$_3$)$_2$-Cl group [65]. The absence of resonances at $\delta\sim4.6$ and $\delta\sim4.8$ ppm, characteristic of terminal unsaturation, also suggests that the arms carry *tert*-Cl end groups.

The functionality of the star was calculated by comparing the integrated peak area of core protons [aromatic ($\delta=6.8$ ppm), -CH$_2$- ($\delta=4.0$ ppm)] to the chain end protons (-CH$_2$- ($\delta=1.95$ ppm), and -CH$_3$ ($\delta=1.65$ ppm)). This procedure gave ~8.1 arms per core after correcting for the presence of ~10% linear contaminant. Thus ¹H NMR spectroscopy provides direct evidence for the formation of the sought *tert*-Cl telechelic octa-arm stars and the number of arms.

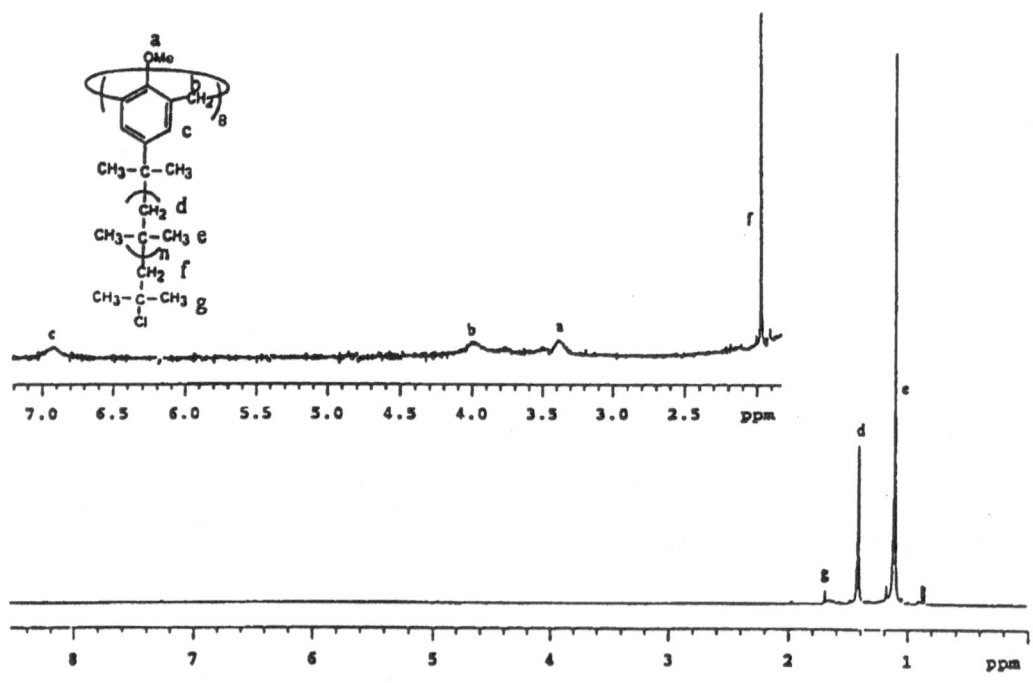

Fig. 3. ¹H NMR (300 MHz) spectrum of *tert*-Cl end-capped octa-arm star $\overline{M}_w=3.76\times10^4$ g/mol) containing a calix[8]arene core

4.2.2
Core Destruction Experiments

To support the ^{1}H NMR evidence, the number of arms in a PIB star was determined by a chemical method. Accordingly, the product of a representative experiment (entry 2; Table 1) was purified (fractionation) and subjected to core destruction. This technique has been repeatedly used in our laboratories to determine the number of arms of stars with aromatic cores and PIB arms [38, 61, 62, 66]. In core destruction the aromatic cores are selectively destroyed by exhaustive oxidation while the saturated aliphatic arms resist oxidation and can be quantitatively determined by GPC (see Scheme 4). Experimentally, it was found that the core was completely destroyed after 16 h under the conditions used (single peak by GPC analysis, see Fig. 4). Control experiments, conducted under similar conditions, showed (by GPC) that calixarenes are destroyed by oxidation [61] and form low molecular weight products, whereas linear PIB survives the oxidation [66].

The molecular weight of surviving arms was determined by GPC (LLS) (see Fig. 4). The dn/dc of the surviving arms was 0.107 cm^3/g. The $\overline{M}_n$, $\overline{M}_w$, and $\overline{M}_w / \overline{M}_n$ were 1.37×10^4 g/mol, 1.50×10^4 g/mol and 1.23, respectively. The dispersity of the arms after core destruction was slightly broader than that of the stars. The number average number of arms, $\overline{N}_{n,arm}$ was determined by dividing the number average molecular weight of the star obtained by GPC (LLS) by the number average molecular weight of the surviving arms obtained by GPC (LLS): $\overline{N}_{n,arm} = (1.05 \times 10^5$ g/mol-1536 g/mol$) \div 1.37 \times 10^4$ g/mol$=7.6$. Similarly, the weight average number of arms, $\overline{N}_{w,arm}$, was determined: $\overline{N}_{w,arm} = (1.16 \times 10^5$ g/mol-1536 g/mol$) \div 1.50 \times 10^4$ g/mol$=7.7$. The number of arms were found to be slightly lower than theoretical (i.e., 8.0) which may be due to incomplete initiation and/or incomplete core oxidation.

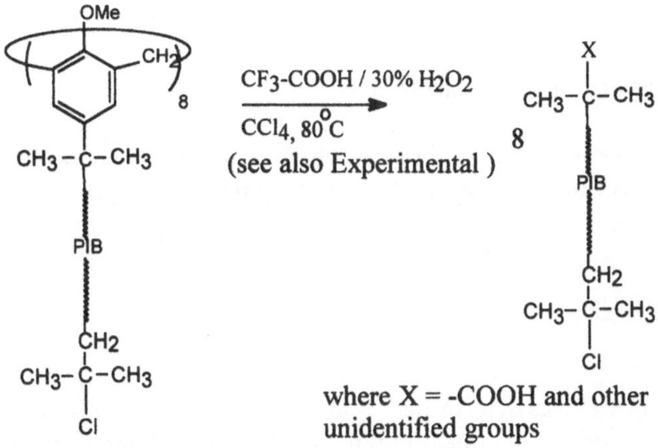

Scheme 4

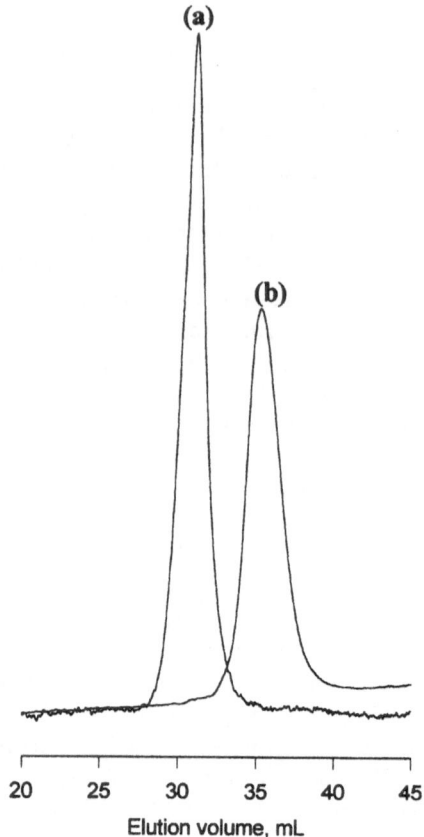

Fig. 4. Gel permeation chromatograms, RI traces: *a* before core destruction; *b* after core destruction of a star polymer (entry 2; Table 1)

4.3
Determination of g* of Octa-Arm PIB Stars

The branching coefficient, g*, of star polymers is defined as the ratio of the radius of gyration of branched and linear polymers having the same molecular weight [67], i.e.,

$$g^* = \frac{\left\langle R_g^2 \right\rangle_{star}}{\left\langle R_g^2 \right\rangle_{linear}} \qquad (1)$$

In this study, g* was determined by using the above equation. The radii of gyration of stars and similar linear polymers [68] were obtained from light scattering measurements. The g* values were calculated as 0.36 and 0.40, respectively, for two samples. The g* values of similar PIB stars were reported to be 0.39 by earlier investigators [54]. Thus the observed g* values also indicate the formation sought multi-arm stars.

In conclusion, well-defined narrow dispersity stars, $\overline{M}_w / \overline{M}_n \sim 1.11$, can be prepared by inducing polymerization with relatively low concentrations of BCl_3 and following up with moderate concentrations of $TiCl_4$. The $BCl_3/TiCl_4$ ratio controls the outcome of the synthesis. Desirable narrow dispersity products formed only when the BCl_3 and $TiCl_4$ coinitiators were added in sequence in two stages and only in the presence of well-chosen BCl_3-$TiCl_4$ concentrations. The sequential addition of BCl_3 and $TiCl_4$ is necessary to obtain desirable product characteristics. Determination of the number of arms by ^{1}H NMR and core destruction indicated close to the theoretical number of arms, i.e., eight.

4.4
Mechanism of Star Synthesis by BCl$_3$-TiCl$_4$ Coinitiators

The experimental observations may be explained by the mechanism shown in Scheme 5 together with the following speculations. The initial event of star synthesis is the formation of 2 from 1 (under the influence of BCl_3 or $TiCl_4$ [7]) which in the presence of excess BCl_3 gives 3. The existence of an equilibrium between dormant and active species such as 2 and 3, respectively, in living IB polymerization has been discussed in detail [7]. The ionicity of the active form 3 and its position in the Winstein spectrum [7] depend on experimental conditions (i.e., nature and concentration of the coinitiator, additives (e.g., organic bases), temperature, solvent polarity, etc.).

During stage I, the active intermediate 3 undergoes slow propagation by route c; to give stars with oligomeric arms. The equilibrium a between the living 3 and dormant 2 is rapid with the dormant species 2 predominating. Thus the symbol 2 stands not only for the initially formed *tert*-benzylic chloride (shown) but also for stars carrying relatively low molecular weight *tert*-chlorine ended PIB arms [i.e., $1(\sim\sim\sim-CH_2-C(CH_3)_2-Cl)_8$]. The reverse reaction of a is very rapid due to the highly unstable $BCl_4^{\ominus}$, and thus deprotonation of 3 or the deprotonation of the low molecular weight stars $1(\sim\sim\sim-CH_2-C(CH_3)_2^{\delta\oplus}--BCl_4^{\delta\ominus})_8$ is absent.

Stage II starts with the introduction of $TiCl_4$ which rapidly converts dormant 2 into active species 4 which in turn undergoes relatively fast propagation by e; the rate of e; is far higher than that of c. Also competing with e is rapid counteranion exchange by path $d[(TiCl_5^{\ominus} + BCl_3 \rightleftharpoons TiCl_4 + BCl_4^{\ominus})$ [69]] which converts 4 via 3 to the dormant form 2.

In the course of stage II, processes b → e → d → a occur in rapid succession and produce the sought stars with *tert*-chlorine-ended PIB arms. During stage II, propagation via c is relatively unimportant. Also, the extent of deprotonation (process f) is negligible because the rate of this process is low relative to processes d and e [69]. Hence BCl_3 serves two purposes: it ensures complete initiation in stage I, and it stabilizes the chain end in stage II by reducing the lifetime of the relatively long-lived $PIB^{\oplus}TiCl_5^{\ominus}$ ion pairs through counteranion ex-

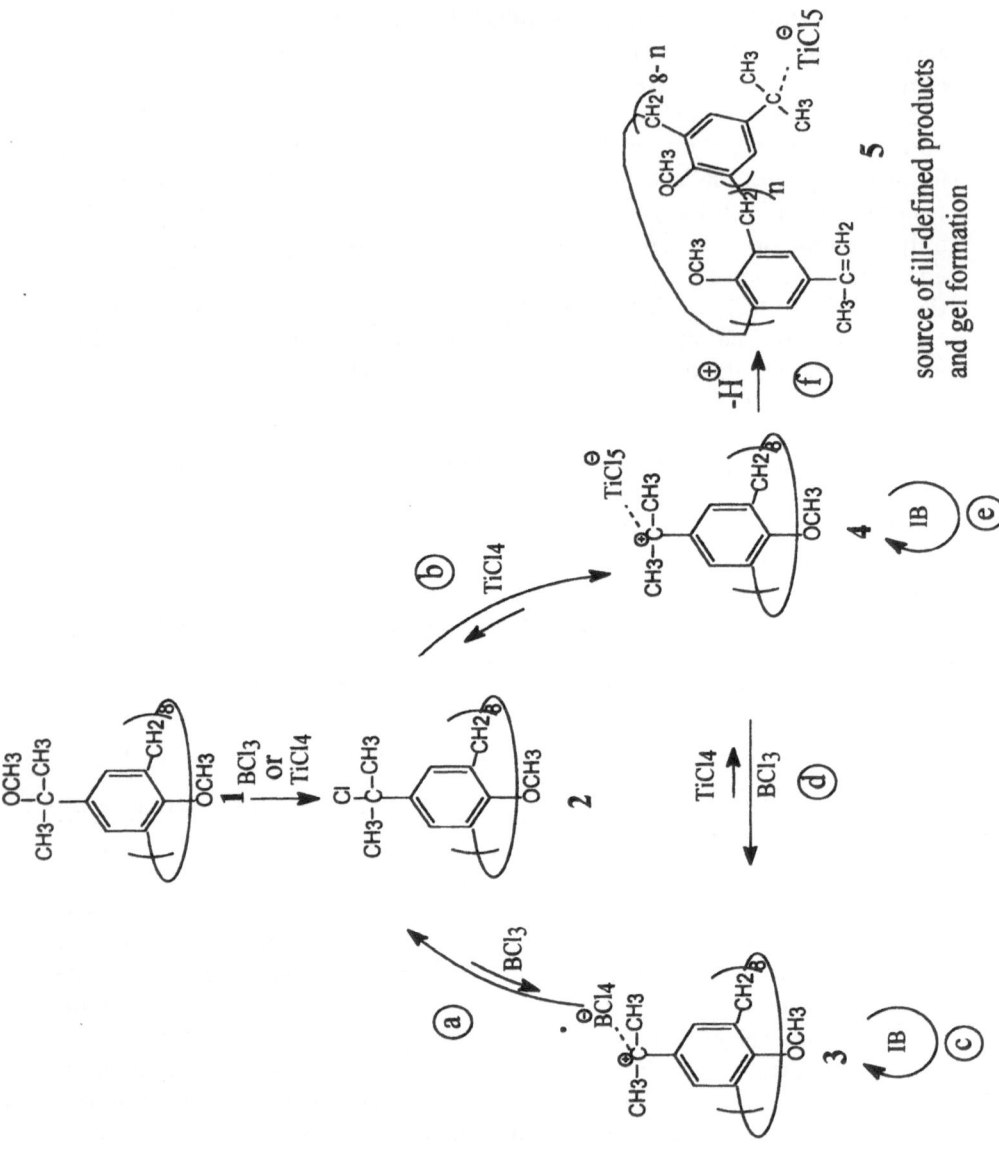

Scheme 5

change, and thus rapidly produces stars with *tert*-Cl end functionalized arms [i.e., 1(~~~~~~-CH$_2$-C(CH$_3$)$_2$-Cl)$_8$].

In contrast, if TiCl$_4$ is used alone the polymerization proceeds only by b $\rightarrow$ e and f. Since TiCl$_5^\ominus$ is more stable than BCl$_4^\ominus$ [69], active intermediate **4** is longer lived than **3**, and will undergo facile proton elimination (process f) to form undesirable structures akin to **5** having α-methylstyrene-type groups. Similar isopropenyl-terminated structures may also arise from growing PIB$^\oplus$ arms by proton elimination (chain transfer). Copolymerization of such double bonds during star formation will lead to high molecular weight stars and/or gel. The broad molecular weight products and/or gels observed even at low IB conversions (~25%) indicate that they form primarily due to the interaction of **5** with growing PIB$^\oplus$ arms during star formation.

5
Synthesis and Characterization of Allyl-Functionalized Octa-Arm PIB Stars

An important advantage of the core-first method is that under living conditions the chain ends are reactive and functionalized stars can easily be prepared by the use of suitable end quenching agents. Functional polymers are technologically important for a variety of applications such as emulsifiers, compatibilizers, adhesion promoters, adhesives, crosslinkers, and coatings [41, 70]. Allyl-functionalized PIB is a useful intermediate for preparation of various other functionalized PIBs [7]. Post-modification of allyl-functionalized octa-arm stars can lead to novel crosslinking agents, additives for coating applications, etc.

The synthesis of allyl-functionalized octa-arm stars was achieved by a 'one-pot' two-step procedure [64]. Scheme 6 shows the steps involved in the synthesis. Living PIB$^\oplus$ stars with a predetermined arm molecular weight were pre-

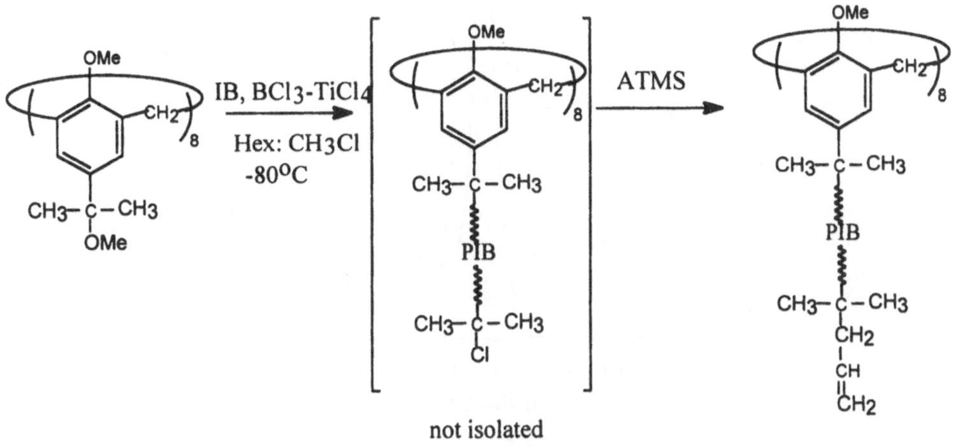

not isolated

Scheme 6

pared by initiator **1** (see Sect. 4.1). The terminal allyl-functionalization of the living PIB$^\oplus$ arms was carried out by a procedure developed in our laboratories [63]. When IB conversions have reached ~95%, ATMS (~100-fold excess relative to the PIB$^\oplus$ chain end) was added to effect allylation.

The stars were characterized by GPC (LLS) and ^{1}H NMR spectroscopy. The RI traces (see Fig. 5) showed the formation of monomodal narrow dispersity stars. Molecular characteristics of two representative samples are summarized in Table 2. According to ^{1}H NMR evidence (Fig. 6), the arms were quantitatively functionalized with allyl groups. The end-functionality of the stars was calculated by comparing the integrated peak area of the core protons [aromatic (δ= 6.82 ppm), -CH$_2$- (δ=4.0 ppm)] and chain end allyl protons (-CH$_2$- (δ=2.0 ppm), -CH= (δ=5.8 ppm), =CH$_2$ (δ=5.1 ppm), and was found to be 8.1, after correcting for the presence of ~10% linear contaminant. Quantitative allyl-functionalization is direct proof for the formation of the octa-arm stars.

In summary, a core-first method was used to prepare allyl-end functionalized octa-arm PIB stars by end quenching the living PIB stars by the use of allyltrimethylsilane. Characterizations by ^{1}H NMR and GPC are consistent with the expected structure and show close to quantitative end-functionalization.

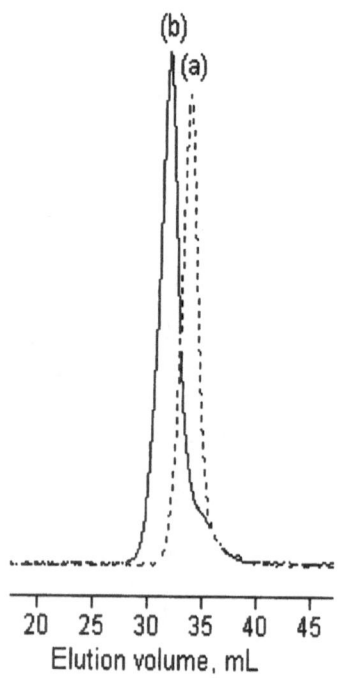

Fig. 5. GPC (RI) traces of two select allyl functionalized octa-arm PIB stars with different molecular weights: *a* $\overline{M}_w$=5.77×10^{4}g/mol; *b* $\overline{M}_w$=11.0×10^{4}g/mol

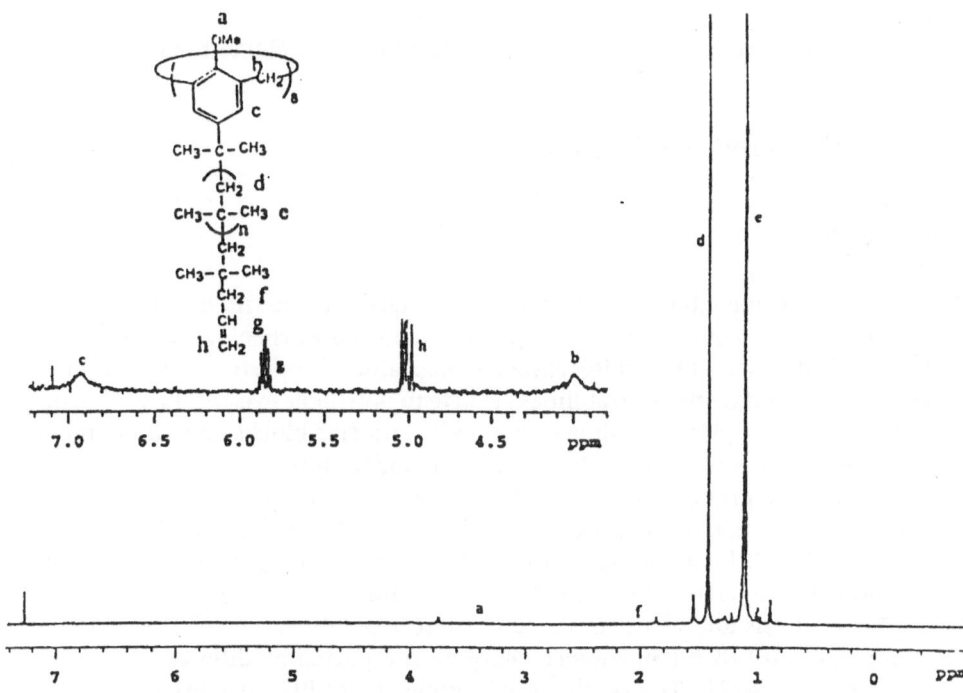

Fig. 6. ^{1}H NMR (600 MHz) spectrum of allyl end-capped PIB star ($\overline{M}_w = 11.0 \times 10^4$ g/mol) containing a calix[8]arene core

Table 2. Molecular characteristics of two select allyl functionalized octa-arm PIB stars

Initiator 1 mol×10^{-5}	Star molecular weight g/mol×10^{-4}				Arm molecular weight[b] g/mol×10^{-3}		End Functionality of Star[c]	
	Observed[a]		Calculated	$\overline{M}_w / \overline{M}_n$	Observed	Calculated	Observed	Corrected
	$\overline{M}_{n(LLS)}$	$\overline{M}_{w(LLS)}$	$\overline{M}_n$		$\overline{M}_{n(LLS)}$	$\overline{M}_n$		
3.52	5.36	5.77	4.94	1.08	6.7	6.2	9	8.1
3.19	9.79	11.1	8.78	1.13	12.2	11.0	8.9	8

[a] Calculated by Astra 4.0; $\overline{M}_n$(LLS) data are meaningful since the molecular weight distributions are narrow
[b] Calculated assuming arm number=8
[c] Calculated from ^{1}H NMR spectroscopy data of virgin samples
[d] Assuming the presence of ~10% linear contaminant

6
Synthesis and Characterization of Novel Thermoplastic Elastomers

6.1
Octa-Arm Poly(styrene-*b*-isobutylene) Stars

6.1.1
Introduction

There is a growing interest in the synthesis of star-block thermoplastic elastomers (TPEs) on account of their unique mechanical and rheological properties [71–73]. PIB-based TPEs exhibit excellent mechanical properties and have superior thermal and oxidative stabilities relative to polydiene-based TPEs [73, 74].

Bi and Fetters [34] have shown that multi-arm star blocks containing polydiene soft-segments prepared by anionic methods exhibit superior mechanical and processing properties compared to the corresponding linear triblocks, especially when arm molecular weights are low. Contrary to what was observed for triblock TPEs [72], sample preparation (molding vs casting) did not much affect the ultimate properties of star blocks with arm numbers higher than six [34].

The synthesis and characterization of linear poly(styrene-*b*-isobutylene-*b*-styrene), PSt-*b*-PIB-*b*-PSt, triblocks have been reported by different groups of investigators [74–77]. The synthesis and properties of PIB-based star-block copolymers with more than three arms [75, 78] have not yet been reported. The syntheses of multi-arm radial star blocks containing PSt-*b*-PIB arms by an arm-first method using divinylbenzene [79, 80] and cyclic siloxanes with Si-H groups [80] as linking agents have been achieved and the properties of the materials are still under investigation.

This section concerns the first synthesis and characterization of novel octa-arm PSt-*b*-PIB star-block copolymers [(PSt-*b*-PIB)$_8$-C8]. The octafunctional initiator 1 was used to prepare PIB stars [61, 62] of desired molecular weight, then St was added sequentially to obtain the sought star blocks. Scheme 6 outlines the synthetic strategy.

6.1.2
Synthesis and Molecular Characterization

The strategy involved initiation of St polymerization by the living PIB$^\oplus$ arms of octa-arm stars. Scheme 7 illustrates the key steps and structures. The overall methodology of living IB polymerization by the two-stage procedure has been followed (see Sect. 4.1). Briefly, stars with predetermined arm molecular weights were prepared by the use of 1, in conjunction with BCl$_3$-TiCl$_4$ coinitiators used in two stages, then St was added to obtain the star blocks.

Optimization studies were carried out with regard to (i) nature of the nonpolar co-solvent (i.e., hexanes vs methylcyclohexane), (ii) TiCl$_4$ concentration, (iii) blocking time, and (iv) St concentration. Intermolecular alkylation occurred

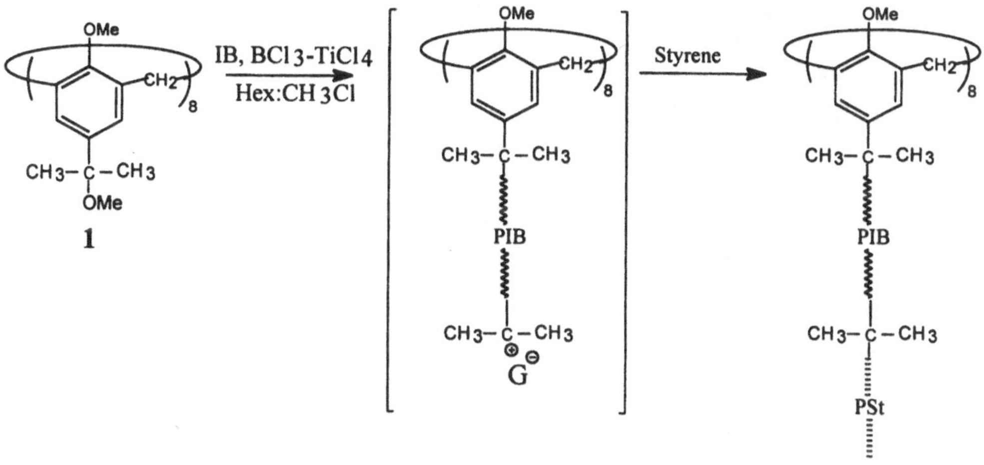

Scheme 7

during PSt block formation in the presence of hexanes as co-solvent, high TiCl$_4$ concentrations (>0.05 mol/l), and with blocking times longer than 60 min [81, 82]. The reduced solubility of star blocks in the CH$_3$Cl:hexanes mixture may lead to aggregation of PSt blocks and promote intermolecular alkylation.

To improve the solubility of the star blocks in the charge, MeCH, was used instead of hexanes (MeCH is a better solvent than hexanes for PSt) [81,82]. In MeCH:CH$_3$Cl (60:40, v/v) mixtures St conversions were relatively higher than in hexanes:CH$_3$Cl and satisfactory star blocks were obtained. To minimize aromatic alkylation and to obtain desirable molecular weight PSt blocks, moderate TiCl$_4$ concentrations (~0.05 mol/l), relatively high St concentrations (~2.5-fold excess), and short blocking times were used. St was added after the IB conversion had reached at least 95%. Stars with varying PSt content were prepared by quenching the polymerization at different times (15–40 min) after the addition of St. Star blocks thus prepared did not show shoulders in the high molecular weight region of GPC traces. Similar to the results obtained in hexanes:CH$_3$Cl solvent mixtures, in the presence of high TiCl$_4$ concentrations or with very long St blocking times, the GPC (RI) traces showed shoulders suggesting the formation of high molecular weight stars, most likely by alkylation of the pendent phenyl rings of PSt by the growing PSt$^\oplus$ [82]. Thus the nature of the co-solvent, TiCl$_4$ concentration in the presence of St, and blocking time after addition of St are critical for molecular weight control and for the preparation of linear (unalkylated, unbranched) PSt segments. The use of too low TiCl$_4$ concentrations produced low molecular weight PSt segments, while the use of too high concentrations of TiCl$_4$ resulted in undesirable alkylation/branching, and at very high concentrations of TiCl$_4$ extensive branching and gelation occurred.

Molecular weight build-up was followed by withdrawing samples from the living charges and analyzing by the GPC. Figure 7 shows the kinetic profile of a

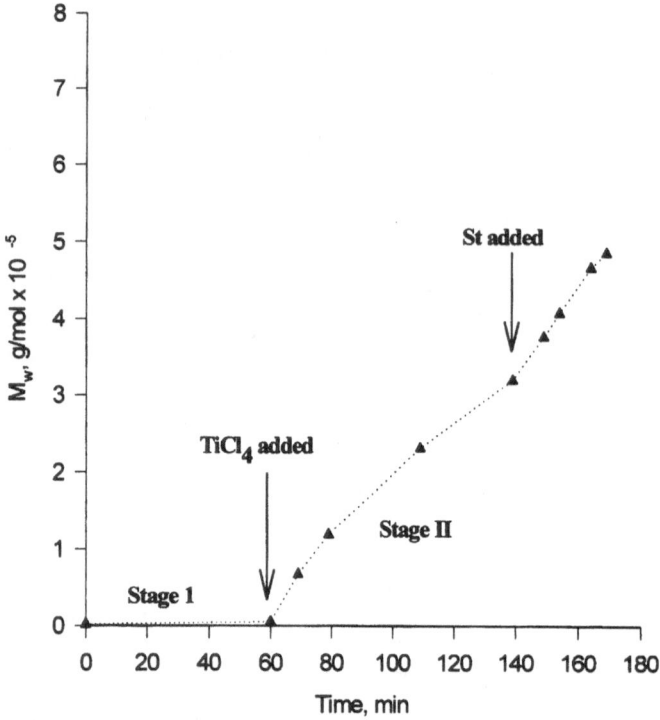

Fig. 7. Molecular weight build-up during star-block formation

representative experiment. During IB polymerization in the presence of BCl_3 (stage I), only a few IB units are added to the initiator, and after the introduction of $TiCl_4$ the molecular weights increased as planned (stage II). After crossover to St (i.e., $PIB^{\oplus}+St \rightarrow PIB\text{-}St^{\oplus}$) the molecular weight increased rapidly because the rate of St polymerization is much higher than that of IB.

Figure 8 shows representative GPC (RI) traces of products obtained before (Fig. 8a) and after (Fig. 8b) St addition. The star block was prepared by two incremental IB additions followed by the addition of St. The crossover from $PIB^{\oplus}$ to $PIB\text{-}St^{\oplus}$ was marked by a color change from pale yellow to deep orange on the addition of St. The charge was homogeneous throughout the polymerization with a gradual viscosity increase after St addition. After quenching and precipitating with methanol, the products were redissolved in MeCH and digested with 5% HCl (to remove coinitiator residues) and methanol. Star blocks thus obtained were readily soluble in THF and were purified by reprecipitating from methanol.

Both GPC traces showed the presence of two products. The major peaks correspond to the sought stars and the minor peaks correspond to by-product(s). The small peak at ~34 ml in Fig. 8a is due to the linear PIB by-product formed by the side reaction discussed in Sect. 4.1. The small peak at ~33 ml in Fig. 8b is

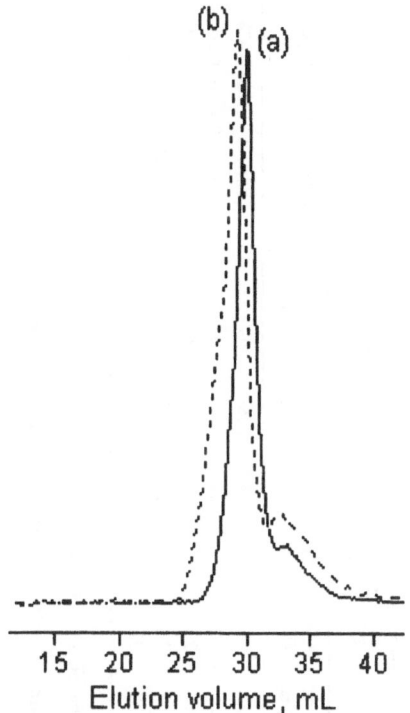

Fig. 8. GPC (RI) traces of products obtained: *a* before St addition; *b* after St addition

a mixture of linear diblocks (PSt-*b*-PIB) formed by crossover of linear living PIB$^{\oplus}$ to St (~10%, by the RI peak area, after extracting the PSt by MEK) plus PSt (3–5%). The peaks corresponding to the stars are monomodal and reflect relatively narrow molecular weight distributions, $\overline{M}_w / \overline{M}_n$ 1.18 and 1.23, respectively.

The molecular weights of the PIB stars and PSt-*b*-PIB star blocks were studied by triple-detector GPC. The RI traces of the star blocks were monomodal, but the molecular weights determined by LLS were higher than expected. This discrepancy may be due to the presence of a small amount of higher molecular weight star blocks formed by intermolecular aromatic alkylation. Thus the molecular weight data of the star blocks and PSt segments calculated from LLS data should be viewed with caution.

Since the presence of small amounts of alkylated (branched) products can distort molecular weights, ^{1}H NMR spectroscopy was used to analyze the PSt content and the molecular weight of the PSt blocks. Molecular characteristics of select samples obtained by GPC, ^{1}H NMR, and conversion are summarized in Table 3. The $\overline{M}_n$ of PIB stars determined by GPC (LLS) were slightly higher than theoretical. This $\overline{M}_n$ and the mole fraction of St obtained by ^{1}H NMR spectroscopy were used to calculate the $\overline{M}_n$ of PSt segments and the PSt content in star

Table 3. Molecular characteristics and compostion of select PSt-b-PIB star blocks

Sample*	PIB segment x 10^{-3} [a] g/mol			PSt-b-PIB star, $\times 10^{-5}$ g/mol						PSt segment $\times 10^{-3}$ $\bar{M}_n$, g/mol			Polystyrene wt%			Molecular structure of stars[e]
	$\bar{M}_n$ (LLS)	$\bar{M}_w$ (LLS)	$\bar{M}_w/\bar{M}_n$	$\bar{M}_n$ (theor)	$\bar{M}_n^{\,b}$ (LLS)	$\bar{M}_n^{\,c}$ (LLS)	$\bar{M}_n^{\,c}$ (LLS)	$\bar{M}_w/\bar{M}_n$	dn/dc cm³/g	GPC (LLS)	^{1}H^d NMR	Conv.	GPC (LLS)	^{1}H^d NMR	Conv.	
19-1	34	40	1.18	32	3.40	3.5	4.1	1.18	0.125	8	7	5.1	19	17	14	(PSt/7-b-PIB/34)$_8$-C8
19-3	34	40	1.18	32	4.34	4.7	5.8	1.23	0.135	20	15	12	37	32	27	(PSt/15-b-PIB/34)$_8$-C8
20-1	25	27	1.10	22	4.30	5.2	6.5	1.26	0.147	29	21	19	54	46	46	(PSt/21-b-PIB/25)$_8$-C8
20-2	34	40	1.18	32	4.48	5.3	6.4	1.21	0.135	22	16	13	39	32	28	(PSt/16-b-PIB/34)$_8$-C8
20-3	46	51	1.12	43	5.56	6.3	7.3	1.18	0.128	22	14	11	32	21	20	(PSt/14-b-PIB/46)$_8$-C8

a Calculated from homo PIB star samples assuming a theoretical number of arms of 8
b Determined by excluding the high molecular weight region: the data were used to calculate the $\bar{M}_n$ of the PSt segments
c Determined by considering the total area
d Samples extracted with MEK. $\bar{M}_n$ of respective PIB segments and mole fraction of St obtained by ^{1}H NMR spectroscopy used. The contribution of the core to the aromatic region has been neglected
e First digit= $\bar{M}_n$ of PSt; second digit= $\bar{M}_n$ of PIB; C8=calix[8]arene

blocks. The $\overline{M}_n$s calculated in this manner may be slightly higher, but more reliable than the values obtained by GPC (LLS). The $\overline{M}_n$ and PSt contents were also calculated from the weight of polymer (conversion) assuming 100% IB conversion. The true molecular weight of the PSt segments is probably between these two values.

6.1.3
Properties of (PSt-b-PIB)$_8$-C8 Stars

6.1.3.1
Thermal Properties

It is known that in PSt-b-PIB-b-PSt triblocks microphase separation starts as the molecular weight of the PSt segments approaches ~5000 g/mol [77]. Such copolymers show two distinct glass transition temperatures (T_g) corresponding to the PIB and PSt phase. The DSC thermogram of a representative star block showed two T_gs corresponding to the PIB (at –67 °C) and PSt (at 104 °C) segments [82]. The thermal degradation behavior of star blocks has been studied by thermogravimetric analysis (TGA). The 5% decomposition temperature, T_d, was 385 °C in N_2 atmosphere [82].

6.1.3.2
Mechanical Properties

The mechanical properties of the star blocks were studied. These star blocks exhibited high tensile strengths (18–26 MPa) and elongations (500–800%). As mentioned above (see Fig. 8) the samples contained 10–15% contamination (diblock or diblock plus PSt). In view of the extraordinary sensitivity of the mechanical properties of linear TPEs to even small (<5%) amounts of diblock contamination, the excellent mechanical properties of star blocks containing a relatively high quantity of diblocks are truly remarkable. The superior mechanical properties of star blocks were analyzed by Bi and Fetters [34]. Figure 9 shows stress-strain traces of select star blocks. The tensile strength increased with an increase in PSt content. The role of PSt domains in the stress absorbing mechanism has been explained [34]. In the case of PSt-b-polydiene-b-PSt TPEs, it has been shown that, on application of stress, failure occurs in the PSt domains [72].

Stars with 32% PSt, (PSt/15-b-PIB/34)$_8$-C8, showed ~26 MPa tensile stress. There was no appreciable difference in the tensile properties of unextracted and MEK extracted star blocks [see (PSt/15-b-PIB/34)$_8$-C8 and (PSt/16-b-PIB/34)$_8$-C8]. The modulus and Shore A hardness were slightly higher for the unextracted star, (PSt/16-b-PIB/34)$_8$-C8, which may be due to the presence of PSt contamination which acts as a rigid filler. A dramatic difference in the tensile behavior was observed when PSt content was increased. Stars with low PIB block molecular weight and high PSt content (46%), e.g., (PSt/21-b-PIB/25)$_8$-C8, showed plastic-like behavior, i.e., it showed a high modulus, a yield point, and a short draw. Ex-

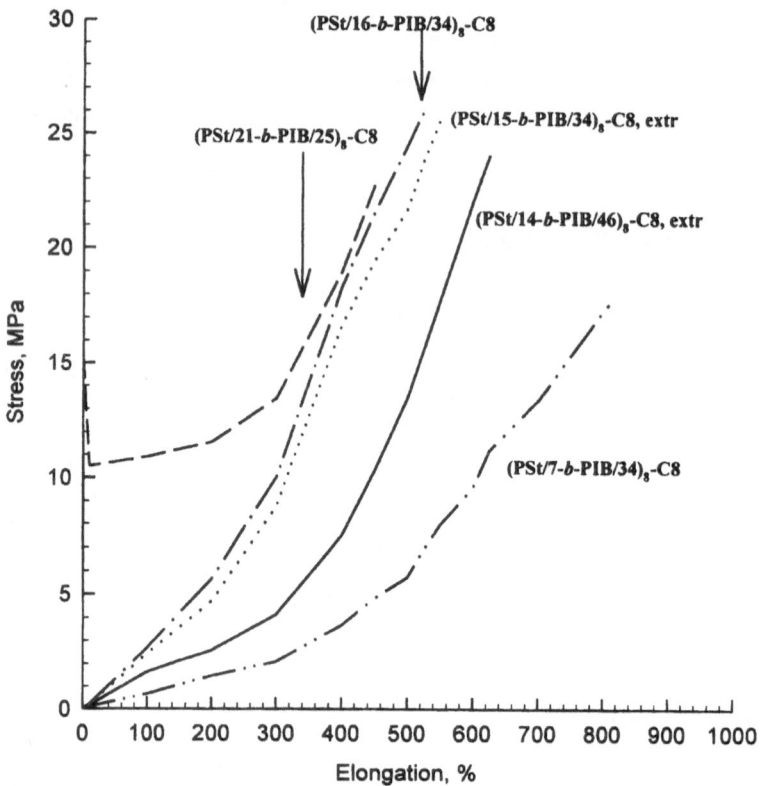

Fig. 9. Stress-strain traces of select star blocks (*numbers in parentheses* indicate $\overline{M}_n \times 10^{-3}$ of segments; C8=calix[8]arene initiator fragment; *extr* MEK extracted)

cept for the star with a relatively high PSt content, (PSt/21-*b*-PIB/25)$_8$-C8, none showed draw behavior. Modulus and hardness showed similar trends, i.e., these values increased with increasing PSt content. Elongation increased with an increase in relative PIB content. The star block with lowest PSt content (~17%) showed the highest elongation (~800%).

Evidently, the products exhibited excellent strengths and elongations (up to 26 MPa and >500%, respectively), in spite of the presence of 10–15% contaminants (PSt and/or PSt-*b*-PIB diblocks) in the samples. The strength of these star blocks is superior to those of the strongest PSt-*b*-PIB-*b*-PSt triblocks reported [77] to date.

Compared to polydiene-based TPEs [34, 72], PIB-based TPEs have somewhat lower strength (~35 MPa vs ~25 MPa). The lower strength of PIB-based TPEs has been postulated to be due to a different failure mechanism, or to the presence of diblock contamination, or to the existence of a diffuse interphase [77]. However, linear PSt-*b*-PIB-*b*-PSt triblock ionomers [78] show higher tensile strength than corresponding linear triblocks, which contradicts the postulate of a different failure mechanism, i.e., failure in the rubbery region. Since the cross-over of PIB$^\oplus$ to PIB-St$^\oplus$ (i.e., crossover from the less reactive IB to the more re-active St) is relatively slow, it is likely that the PSt blocks formed are of a relative-ly broader molecular weight distribution. This could lead to a diffuse interphase and to the formation of irregular domains upon microphase separation with a consequent reduction in strength. More results are needed to elucidate these postulates.

6.1.3.3
Dynamic Melt Viscosity Studies

The dynamic melt viscosity measurements of select star blocks and a similar tri-block were carried out on a rheometric mechanical spectrometer, RMS. Circular molded samples of ~2 cm diameter and ~1.5 mm thickness were subjected to forced sinusoidal oscillations. Dynamic viscosities were recorded in the fre-quency range of 0.01–100 rad/s at 180 °C. Figure 10 shows the complex viscosi-ties of two select star blocks and a similar linear triblock. The plots showed char-acteristic behavior of thermoplastic elastomers, i.e., absence of Newtonian be-havior even in the low frequency region. The complex viscosity of the star block

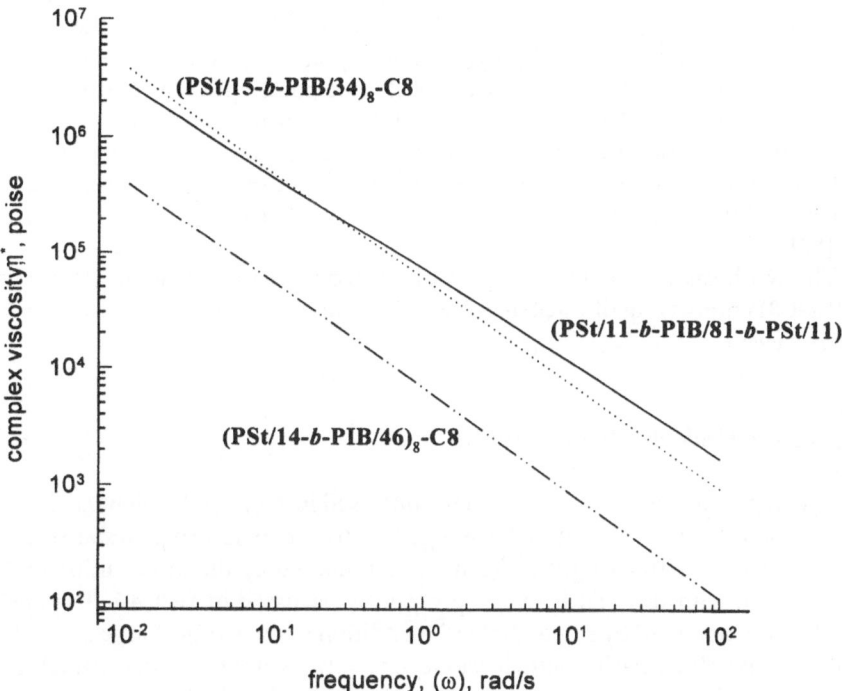

Fig. 10. Dynamic viscosity of two select PSt-b-PIB star blocks and a triblock at 180 °C

(PSt/14-b-PIB/46)$_8$-C8 with 21% PSt content was more than a decade lower than the linear triblock with 20% PSt content. The star block with 32% PSt content, (PSt/16-b-PIB/34)$_8$-C8, showed a higher melt viscosity than the one containing 21% PSt, PSt/14-b-PIB/46)$_8$-C8, and the values are comparable to that of the triblock with a much lower molecular weight. These results indicate that the star blocks show better flow properties than triblocks at comparable PSt contents. It is known that the melt viscosities of TPEs depend mainly on relative PSt content and PSt block length [83, 84]. Leblanc [85] observed that the activation energy for flow for poly(styrene-b-butadiene) stars was lower than that of similar linear triblocks with comparable PSt content. The lower activation energy indicated an easier flow mechanism and it was hypothesized that aggregates of stars flow [85].

6.2
Octa-Arm Poly(p-chlorostyrene-b-isobutylene) Stars

6.2.1
Introduction

During the synthesis of octa-arm stars with poly(styrene-b-isobutylene) arms intermolecular aromatic alkylation also occurred as a minor side reaction during polystyrene (PSt) block formation [81, 82]. Such reactions lead to branching of stars and to products with high molecular weight. One way to eliminate this problem is to use a para-protected styrene, e.g., p-chlorostyrene (pClSt) instead of styrene. Moreover, the T_g of poly(p-chlorostyrene), PpClSt, is higher than that of PSt, T_g, PpClSt=129 °C [86]. Also, PpClSt has good optical properties and is flame resistant [87]. The surface properties of such thermoplastic elastomers, TPEs, could be different from PSt containing TPEs on account of the presence of polar PpClSt segments. The synthesis and properties of PIB-b-PpClSt-b-PSt triblocks and three-arm star blocks with PpClSt-b-PIB arms have been investigated [88].

The synthesis and characterization of novel octa-arm star blocks ((PpClSt-b-PIB)$_8$-C8) comprising of PpClSt-b-PIB arms radiating from a calix[8]arene core are reviewed.

6.2.2
Synthesis and Molecular Characterization

The synthesis strategy was similar to the one used in Sect. 6.1.2 which involved initiation of pClSt polymerization by living PIB$^{\oplus}$ arms of octa-arm PIB stars [89, 90]. Octa-arm stars with living PIB$^{\oplus}$ arms were prepared by the use of initiator 1 in conjunction with BCl$_3$-TiCl$_4$ by the procedure discussed in Sect. 4.1. The PpClSt blocks were obtained by sequential pClSt addition to the living charge at ~95% IB conversions [89]. CH$_3$Cl: methylcyclohexane solvent mixtures were suitable for the living polymerization of pClSt [87, 88]. Hence blocking was carried out in CH$_3$Cl:methylcyclohexane mixtures, 40:60 (v/v) with TiCl$_4$=0.057 mol/l, and by the use of a twofold excess of pClSt relative to the targeted molecular weight of Pp-ClSt at −80 °C. The crossover from PIB$^{\oplus}$ to PIB-pClSt$^{\oplus}$ was marked by a color change from pale yellow to deep orange on pClSt addition. The charge was homogeneous throughout the polymerization. A slow increase in the viscosity of the charge was noted after the addition of pClSt. The star blocks were readily soluble in THF and were purified by reprecipitating into methanol. According to pClSt conversions, only a short PpClSt block ($\overline{M}_n$=4700 g/mol) was formed after 60 min. To obtain longer PpClSt blocks, the blocking time was increased to 150 min. Products thus obtained contained ~20% PpClSt. Star blocks containing more than 20% PpClSt were prepared under slightly more polar conditions (CH$_3$Cl:methylcyclohexane, 50:50) and higher TiCl$_4$ concentrations (0.063 mol/l). The pClSt conversions and the molecular weights of the PpClSt blocks depend on solvent polarity, TiCl$_4$ and pClSt concentrations, and blocking time.

Figure 11 shows representative GPC (RI) traces of products obtained before and after pClSt addition. The sharp peaks at ~30 and ~29 ml, respectively, correspond to (PIB)$_8$-C8 and (PpClSt-b-PIB)$_8$-C8 and the minor peaks are due to side products. The small peak (~12% by RI peak area) at ~34 ml is due to the lin-

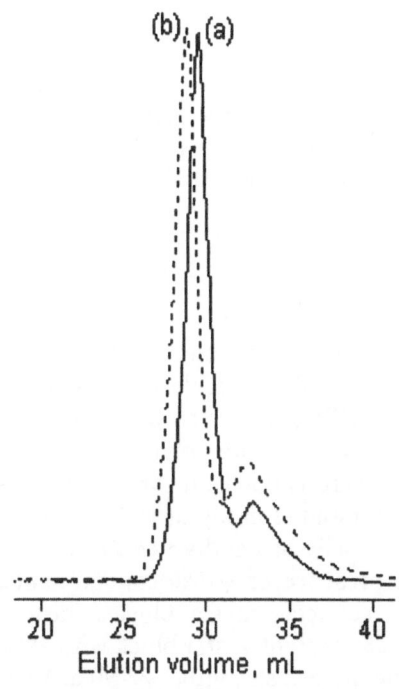

Fig. 11. GPC (RI) trace of products: a before pClSt addition ((PIB)$_8$-C8); b after pClSt addition [(PpClSt-b-PIB)$_8$-C8]

ear PIB by-product [61, 62]. The other small peak at ~33 ml is due to a mixture of linear diblocks (PpClSt-b-PIB) formed by the crossover of linear living PIB$^{\oplus}$ to pClSt (~12%, by RI peak area, after extracting the PpClSt with MEK) and Pp-ClSt (4–6%). The sharp peak corresponding to the star block is monomodal and reflects relatively narrow molecular weight distributions ($\overline{M}_w$ / $\overline{M}_n$=1.13) and shoulders are absent even after long blocking times. These observations were significantly different from those observed during the blocking of St [81, 82]. It is noteworthy that the dispersity of the star block did not increase after pClSt addition, whereas in the case of blocking with St, the star block always showed broader dispersity than the corresponding star-PIBs [81, 82]. According to these observations, intermolecular alkylation was absent.

Molecular weights of PpClSt-b-PIB stars were determined by GPC equipped with a light scattering detector. The PpClSt content was also analyzed by ^{1}H NMR spectroscopy and from total conversions assuming 100% conversion for IB. The dn/dc values and molecular characteristics of select star blocks are summarized in Table 4. The molecular weights and PpClSt contents determined by GPC (LLS) were in good agreement with the values calculated by ^{1}H NMR spectroscopy and conversions.

6.2.3
Physical Properties of (PpClSt-b-PIB)$_8$-C8 Stars

6.2.3.1
Thermal Properties

The DSC thermogram of a representative star block showed two T$_g$s corresponding to the PIB (at –67 ˚C) and PpClSt (at 127 ˚C) segments. The star block showed high thermal stability. The 5% decomposition temperature, T$_d$, was 390 ˚C in N$_2$.

6.2.3.2
Mechanical Properties

The mechanical properties of the star blocks were analyzed by tensile measurements. As mentioned earlier, all star blocks contained 15–17% contaminant (diblock or diblock plus PpClSt). Stress-strain measurements were performed without removing these contaminants. Polymer sheets with thickness of ~1 mm were molded and tests were carried out on dumbbell-shaped samples. Thin transparent sheets were molded easily at ~175 ˚C and ~6000 psi. These star blocks exhibited high tensile strengths (22–27 MPa) and elongations (400–650%) in spite of the presence of ~15% contaminants. Figure 12 shows the stress-strain traces of select octa-arm star blocks. The tensile strength increased with an increase in PpClSt content. A star block with 29% PpClSt, (PpClSt/16-b-PIB/38)$_8$-C8, showed the highest strength, ~27 MPa. A star block with relatively high PpClSt content (51%), (PpClSt/19-b-PIB/18)$_8$-C8, behaved like a plastic. It showed a high modulus, a yield point, a short elastic extension, and break.

Table 4. [a]Molecular characteristics and compostion of some select octa-arm PpClSt-b-PIB star copolymers

Sample	IB polymerization			Blocking				PpClSt wt%		Conv.	Composition[d]
	PIB star $\bar{M}_w$ $\times 10^{-5}$ g/mol	PIB arm[b] $\bar{M}_w$ $\times 10^{-3}$ g/mol	$\bar{M}_w/\bar{M}_n$	dn/dc cm^3/g	Block star $\bar{M}_w$ $\times 10^{-5}$ g/mol	$\bar{M}_w/\bar{M}_n$	PpClSt arm[b] $\bar{M}_w$ $\times 10^{-3}$ g/mol	GPC	^{1}H NMR[c]		
18–3	3.10	38	1.12	0.130	3.99	1.13	11	22	23	21	(PpClSt/11-b-PIB/38)$_8$-C8
21–1	1.42	18	1.09	0.141	2.90	1.14	19	51	52	52	(PpClSt/19-b-PIB/18)$_8$-C8
21–2	3.15	39	1.13	0.135	4.45	1.13	16	29	31	30	(PpClSt/16-b-PIB/39)$_8$-C8
21–3	3.66	46	1.12	0.132	4.86	1.12	15	25	27	23	(PpClSt/15-b-PIB/46)$_8$-C8

[a] Molecular weight and dispersity determined by LLS. Astra 4.0
[b] Calculated from homo PIB star based on theoretical number of arms, 8
[c] ^{1}H NMR spectroscopy measurements were made on samples extracted with MEK. Contribution of the core to the aromatic region has been neglected
[d] First digit= $\bar{M}_w$ of PpClSt; second digit= $\bar{M}_w$ of PIB; C8=calix[8]arene

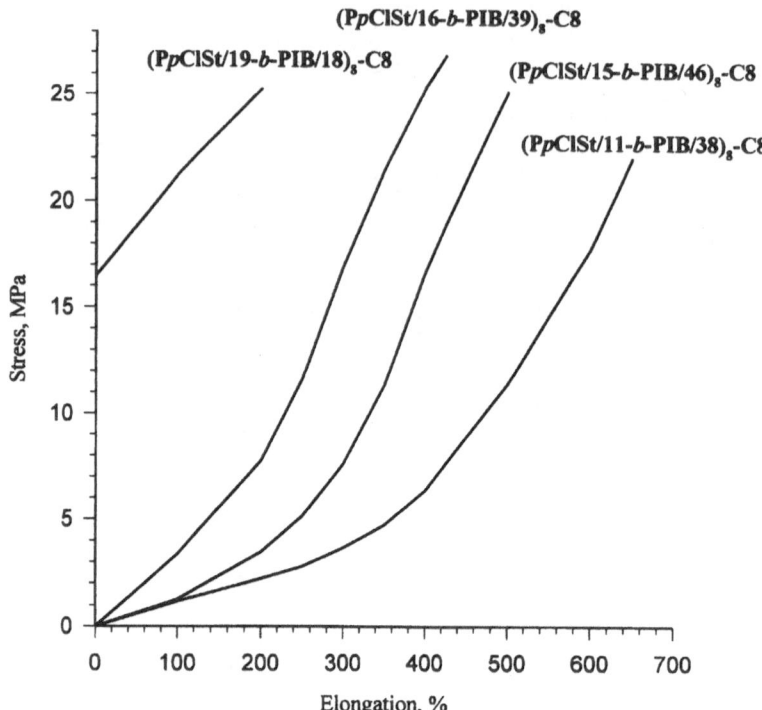

Fig. 12. Stress-strain traces of select star blocks (*numbers in parentheses* indicate $\overline{M}_n \times 10^{-3}$ of segments; C8=calix[8]arene initiator fragment)

These star block TPEs show higher tensile strengths and moduli, and lower ultimate elongations, than similar stars with PSt-*b*-PIB arms, which may be due to the presence of the high T_g PpClSt blocks. The tensile strengths of these star blocks are superior to similar three-arm star blocks and triblock TPEs [88].

6.2.3.3
Morphology

Figure 13 shows the TEM micrograph of a star block containing 22% PpClSt. The morphology shows phase-separated domains of PpClSt and PIB. The PpClSt formed spherical domains with diffused interphase and are irregularly dispersed in the PIB matrix. The formation of irregular domains with diffused interphase is attributed to the presence of diblock and PpClSt contaminants.

In conclusion, the synthesis of novel stars comprising eight PSt-*b*-PIB and Pp-ClSt-*b*-PIB arms radiating from a calix[8]arene core has been accomplished. Overall compositions and relative molecular weights of the glassy segments can be controlled by suitable choice of blocking conditions such as solvent polarity,

Fig. 13. TEM micrograph of star block 18–3 (22% PpClSt content)

TiCl$_4$ concentration, and blocking time. The star blocks exhibit an excellent combination of thermoplastic elastomer properties. The products exhibited excellent strengths and elongations (up to 27 MPa and ~500%), in spite of the presence of 10–15% contaminants. They are potential easily processiable TPEs.

7
Conclusions

This review concerns the synthesis and characterization of well-defined novel octa-arm star PIBs, octa-arm allyl-telechelic PIB stars, and octa-arm star-block copolymers (PIB-b-PSt and PIB-b-PpClSt) wherein the arms emanate from a calix[8]arene core. The syntheses were accomplished by the use of a novel octa-functional initiator. The shear and oxidative stability of PIB arms is high so that these stars may be of use as crankcase additives. Allyl-functionalized PIB stars may be suitable for the preparation of coatings. The star blocks show excellent thermoplastic elastomeric properties, with high strength and elongation. The flow properties of the star blocks are superior to corresponding triblocks with similar PSt content. All star blocks were easily moldable (~170 °C and ~6000 psi). By the judicious choice of glassy-block lengths, star thermoplastic elastomers with superior strength and processability can be prepared.

Acknowledgment. This material is based on work supported by the NSF under Grant DMR-94–23202.

8
References

1. Roovers J (1985) In: Kroschwitz JI (ed) Encyclopedia of polymer science and engineering, vol 2, 2nd edn. Wiley-Interscience, New York, p 478
2. Source-based nomenclature for copolymers (1985) Pure Appl Chem 57:1427
3. Bywater S (1979) Adv Polym Sci 30:90
4. Worsfold DJ, Zilliox JG, Rempp P (1969) Can J Chem 47:3379
5. Bauer BJ, Fetters LJ (1978) Rubber Chem Technol 51:406
6. (a) Hsieh HL, Quirk RP (1996) Anionic polymerization principles and practical applications. Marcel Dekker, New York, p 333; (b) McGrath JE (1981) In: Anionic polymerization: kinetics, mechanism and synthesis. ACS Symposium Series No 166, Washington, DC
7. Kennedy JP, Ivan B (1992) Designed polymers by carbocationic macromolecular engineering: theory and practice. Hanser, Munich
8. Morton M, Helminiak TE, Gadkary SD, Bueche F (1962) J Polym Sci 57:471
9. Sogah DY, Hertler WR, Webster OW, Cohen GM (1987) Macromolecules 20:1473
10. Kuriyama A, Otsu T (1984) Polym J 16(6):511
11. Gaynar SG, Matyjaszewski K (1997) Polym Prepr Am Chem Soc Div Polym Chem 38(1):758
12. Nagasawa M, Fujimoto T (1972) In: Okamura S, Takayanagi M (eds) Progress in polymer science Japan. Wiley, New York, p 278
13. Fijimoto T, Tani S, Takano K, Ogawa M, Nagasawa M (1978) Macromolecules 11:673
14. Gordon B, Blumenthal M, Loftus JE (1984) Polym Bull 11:349
15. Eschwey H, Hallensleben ML, Burchard W (1973) Makromol Chem 173:235
16. Gnanou Y, Lutz P, Rempp P (1988) Makromol Chem 189:2885
17. Quirk RP, Yoo T, Lee B (1994) J Macromol Sci, Pure Appl Chem A(31)8:911
18. Mishra MK, Wang B, Kennedy JP (1987) Polym Bull 17:307
19. Chen CC, Kaszas G, Puskas JE, Kennedy JP (1989) Polym Bull 22:463
20. Ivan B, Kennedy JP (1988) Polym Mater Sci Eng 58:866
21. Huang KJ, Zsuga M, Kennedy JP (1988) Polym Bull 19:43
22. Shohi H, Sawamoto M, Higashimura T (1991) Macromolecules 24:4926
23. Shohi H, Sawamoto H, Sawamoto M, Fukui H, Higashimura T (1994) J Macromol Sci, Pure Appl Chem A31(11):1609
24. Franta E, Reibel L, Lehmann J, Penczek S (1976) J Polym Sci, Symp 56:139
25. Schultz RC (1993) Makromol Chem, Macromol Symp73:103
26. Chang JY, Ji HJ, Han MJ (1994) Macromolecules 27:1376
27. Cloutet E, Fillaut J, Gnanou Y, Astruc D (1994) J Chem Soc, Chem Commun 2433
28. Zhu Z, Rider J, Yang CY, Gilmartin ME, Wnek GE (1992) Macromolecules 25:7330
29. Menceloglu YZ, Baysal BM (1992) Angew Makromol Chem 200:37
30. Schaefgen JR, Flory PJ (1948) J Am Chem Soc 70:2709
31. Kricheldorf HR, Adebahr T (1993) Makromol Chem 194:2103
32. Worsfold DJ (1970) Macromolecules 3:514
33. Young RN, Fetters LJ (1978) Macromolecules 11:899
34. Bi LK, Fetters LJ (1976) Macromolecules 9:732
35. Higashimura T, Sawamoto M, Kanaoka S (1991) Macromolecules 24:2309
36. Higashimura T, Sawamoto M, Kanaoka S (1992) Macromolecules 25:6414
37. Kennedy JP, Marsalko TM, Majoros I (1995) US Pat 5,395,885
38. Marsalko TM, Majoros I, Kennedy JP (1993) Polym Bull 31:665
39. (a) Storey RF, Shoemake KA, Chishom BJ (1996) J Polym Sci, Part A Polym Chem 34:2003; (b) Storey RF, Shoemake KA (1998) J Polym Sci, Part A Polym Chem 36:471
40. Asthana S, Majoros I, Kennedy JP (1997) Polym Mat Sci Eng 77:187
41. Simms JA (1991) Rubber Chem Technol 64:139

42. Hadjichristidis N, Tselikas G, Efstratiadis V, Mays JW, Yunan W, Li J (1994) Polym Int 33:171
43. Roovers JEL, Bywater S (1972) Macromolecules 5:384
44. Hadjichristidis N, Roovers JEL (1974) J Polym Sci Polym Phys Ed 12:2521
45. Hadjichristidis N, Fetters LJ (1980) Macromolecules 13:191
46. Zhou LL, Roovers J (1993) Macromolecules 26:963
47. Thomas EL, Alward DB, Kinning DJ, Martin DC, Handlin DL Jr, Fetters LJ (1986) Macromolecules 19:2197
48. Kinning DJ, Thomas EL, Alward DB, Fetters LJ, Handlin DL Jr (1986) Macromolecules 19:1288
49. Pennisi RW, Fetters LJ (1988) Macromolecules 21:1094
50. Iatrou H, Hadjishristidis N (1992) Macromolecules 25:4649
51. Webster OW (1990) Makromol Chem, Macromol Symp 33:133
52. Fukui H, Sawamoto M, Higashimura T (1993) J Polym Sci, Part A Polym Chem Ed 31:1531
53. Fukui H, Sawamoto M, Higashimura T (1994) Macromolecules 27:1297
54. Omura N, Kennedy JP (1997) Macromolecules 30:3204
55. Shim JS (1997) The University of Akron, pers. comm.
56. Pinnazzi C, Esnault J, Lescuyer G, Villette JP, Pleurdeau A (1974) Makromol Chem 175:705
57. Zhou G, Smid J (1993) Polymer 34:5128
58. Gutsche CD (1989) Calixarenes. The Royal Society of Chemistry, Thomas Graham House, Cambridge, p 21
59. Gutsche CD, Dhawan B, No KH, Muthukrishnan R (1981) J Am Chem Soc 103:3782
60. Lin LG, Gutsche CD (1986) Tetrahedron 42:1633
61. Jacob S, Majoros I, Kennedy JP (1996) Macromolecules 29:8631
62. Jacob S, Majoros I, Kennedy JP (1997) Polym Mat Sci Eng 76:298
63. Wilczek L, Kennedy JP (1987) J Polym Sci, Part A Polym Chem 25:3255
64. Jacob S, Majoros I, Kennedy JP (1997) Polym Bull (in press)
65. Si J, Kennedy JP (1994) J Polym Sci, Part A Polym Chem 32:2011
66. Kennedy JP, Ross LR, Nuyken O (1981) Polym Bull 5:5
67. Zimm BH, Stockmayer WH (1949) J Chem Phys 17:1301
68. Fenyvesi G (1997) pers. comm.
69. Pernecker T, Kennedy JP (1994) Polym Bull 33:13
70. Patil AO (1997) Polym Mat Sci Eng 76:85
71. (a) Legge NR, Davison S, De La Mare HE, Holden G, Martin MK (1985) In: Tess RW, Poehlin GW (eds) Applied polymer science. American Chemical Society, Washington, DC, p 175; (b) Noshay A, McGrath JE (1977) In: Block copolymer: overview and critical survey. Academic Press, New York
72. Quirk RP, Morton M (1996) In: Holden G, Legge NR, Schroeder HE (eds) Thermoplastic elastomers, 2nd edn. Hanser, Munich, p 71
73. Puskas JE, Kaszas G (1996) Rubber Chem Technol 69:462
74. Storey RF, Chisholm BJ, Lee Y (1993) Polymer 34:4330
75. Kaszas G, Puskas JE, Kennedy JP, Hager WG (1991) J Polym Sci, Part A Polym Chem 29:427
76. Storey RF, Chisholm BJ, Choate KR (1994) J Macromol Sci, Pure Appl Chem A31(8):969
77. Gyor M, Fodor Z, Wang HC, Faust R (1994) J Macromol Sci, Pure Appl Chem A31(12):2055
78. Storey RF, Chisholm BJ, Lee Y (1992) Polym Prepr Am Chem Soc Div Polym Chem 33:184
79. Storey RF, Shoemake KA (1996) Polym Prepr Am Chem Soc Div Polym Chem 37(2):321
80. Shim JS, Asthana S, Omura N, Kennedy JP (1998) Polym Prepr Am Chem Soc Div Polym Chem 39(1):196

81. Jacob S, Majoros I, Kennedy JP (1997) Polym Mat Sci Eng 77:185
82. Jacob S, Majoros I, Kennedy JP (1998) Rubber Chem Technol (in press)
83. Arnold KR, Meier DJ (1970) J Appl Polym Sci 14:427
84. Kraus G, Naylor FE, Rollmann KW (1971) J Polym Sci Part A-2 9:1839
85. Leblanc JL (1976) Rheol Acta 15:654
86. Brandrup J, Immergut EH (1989) Polymer handbook, 3rd edn. Wiley-Interscience, New York
87. Kennedy JP, Kurian J (1990) Macromolecules 23:3736
88. Kennedy JP, Kurian J (1990) J Polym Sci, Part A Polym Chem 28:3725
89. Jacob S, Majoros I, Kennedy JP (1998) Polym Prepr Am Chem Soc Div Polym Chem 39(1):198
90. Jacob S, Kennedy JP (1998) Polym Bull 41:164

Editor: Prof. J. E. McGrath
Received: July 1998

Polymerization by Metalloporphyrin and Related Complexes

Hiroshi Sugimoto, Shohei Inoue

Department of Industrial Chemistry, Faculty of Engineering, Science University of Tokyo, 1–3 Kagurazaka, Shinjuku-ku, Tokyo 162–8601, Japan
e-mail: hrssgmt@ci.kagu.sut.ac.jp

This review describes the recent development of the living and immortal polymerizations of various monomers controlled by using as excellent initiators the metalloporphyrins of aluminum, zinc, manganese, and tin. Aluminum porphyrin initiates the living polymerizations of methacrylic esters, acrylic esters, and methacrylonitirile, and the living and immortal polymerizations of epxides, and lactones, affording polymers of very narrow molecular-weight distribution, where the rates of polymerization are much accelerated by the presence of an appropriate Lewis acid as an electrophilic monomer activator. Zinc N-alkylporphyrin brings about the living and immortal polymerizations of epoxides and episulfides. Manganese porphyrin also gives polyethers of a fairly narrow molecular-weight distribution by the living and immortal polymerization of epoxides. Tin porphyrin produces poly(methyl methacrylate) via a radical mechanism.

Keywords: Anionic polymerization, Living Polymerization, Immortal polymerization, Metalloporphyrin, Lewis acid

Advances in Polymer Science, Vol.146
© Springer-Verlag Berlin Heidelberg 1999

1
Introduction

The chemistry of metalloporphyrins and related compounds has been developed, especially in the last few decades, in connection with their prominent functions and properties in vital systems [1–5]. For example, iron porphyrin, heme, is known as the active site of hemoproteins such as hemoglobin and myoglobin, in which heme exhibits important biological function such as reversible oxygen binding for transport and storage [6,7]. Iron porphyrins in cytochromes serve as catalysts for oxygenation [8]. Metalloporphyrins of magnesium and cobalt are also of much interest. Magnesium complexes of chlorins, dihydroporphyrins, which are found in chlorophyll, play an important role in biological photosynthesis, the fixation of carbon dioxide by green plants and some microorganisms [9]. Metalloporphyrin derivatives in vitamin B_{12} are of cobalt [10]. Therefore, the attention of the chemists in this field has been focused on mimicking the oxygen-binding property, the electron-transporting ability, and the enzymatic catalysis of hemoproteins in bioorganisms by using artificial metalloporphyrin systems, and a lot of reports have been presented to provide a better understanding of their vital functions [11,12]. However, from the view point of synthetic organic chemistry, most of these studies have been limited to dealing with the catalysis of metalloporphyrins in oxidation or in electron transfer to mimic the biological process. In addition, the potential utility of metalloporphyrins as catalysts for controlling synthetic organic reactions has received little interest, although a wide variety of porphyrins and metalloporphyrins are now easily synthesized thanks to the progress in the synthetic procedures.

The present review describes the novel approaches in metalloporphyrin chemistry to regulating the reactions at the central metal–axial ligand bond in five-and/or six-coordinated complexes. The feature of metalloporphyrins is of course the widely conjugated, rigid and planar macrocyclic ligand called "porphyrin", which invests the central metal with an unusual circumstance, sterically as well as electronically. Therefore, when using metalloporphyrins as catalysts in organic reactions, the reactivity of the metalloporphyrin is expected to be controlled and enhanced by the following factors; (i) site-isolation by virtue of the presence of the widely conjugated, rigid and planar macrocyclic ligand, (ii) excitation of the porphyrin ring by visible light, (iii) structure of peripheral substituents of porphyrin, and (iv) a trans effect of external ligands coordinating to the Lewis acidic central metal.

2
Polymerization with Aluminum Porphyrin

2.1
Living Polymerization of Methacrylic Ester with
Aluminum Porphyrin–Organoaluminum Compound Systems

One characteristic of polymerizations initiated with aluminum porphyrin is the absence of side reactions. A good example is the living polymerization of methacrylic esters. As a representative type of initiator, an organolithium system usually requires a very low reaction temperature to obtain the polymer with a satisfactorily narrow distribution of molecular weight. A higher temperature tends to cause side reactions. In contrast, a polymer with a narrow molecular-weight distribution can be obtained even at room temperature with aluminum porphyrin as the initiator. The absence of side reactions stems from the mild reactivity of aluminum porphyrin. On the other hand, mild reactivity often results in a relatively slow rate of reaction, which is disadvantageous for practical purposes. In fact, the living polymerization of methyl methacrylate (MMA) with aluminum porphyrin is not fast enough. Of course, it is rather ridiculous to lower the reaction temperature to suppress side reactions for a system with high reactivity. In synthetic procedures, however, controlling the reaction to develop selectivity is often a matter of compromise, forcing the chemist to settle for less favorable reaction conditions. Therefore, it is best to design a system based on an amphiphilic bimetallic system that enables us to control the reaction without sacrificing reaction conditions, and then we would expect that appropriate Lewis acids would help aluminum porphyrin to realize accelerated reactions by coordinative activation of the monomer.

We have developed new initiators based on aluminum complexes of tetraphenylporphyrin which can be applied to various types of polymerization reactions, such as ring-opening polymerizations of epoxide [13] and lactone [14], and addition polymerizations of acrylic monomers [15,16]. During the course of this study we recently discovered a new method, "high-speed living polymerization", by using aluminum porphyrin as a nucleophilic initiator in conjunction with an organoaluminum [17,18] or organoboron compound [19] as an electrophilic monomer activator (Lewis acid). This is the first successful example of a clean anionic polymerization assisted by an externally added Lewis acid, which enables us to synthesize uniform molecular-weight polymers within only a few seconds [17]. The basic concept of high-speed living polymerization involves the coordinative activation of a monomer by a Lewis acid. Although such a methodology has been utilized in the field of organic synthesis, an undesired direct reaction between nucleophile and electrophile always occurs in competition with the main reaction. Therefore, of primary importance for the application of this concept to living polymerization is how to suppress undesired side reactions. To date, three different approaches have been developed in order to realize the coexistence of the nucleophilic growing species and a Lewis acid without degrada-

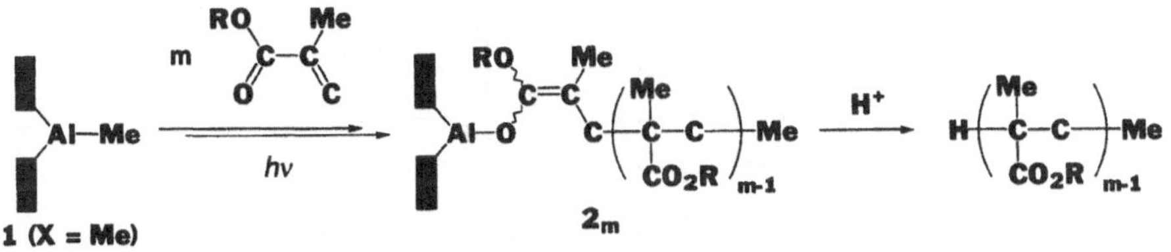

Scheme 1

tive neutralization leading to the termination of polymerization. The first one is to carry out the polymerization reaction thermostated at a low temperature. The second is to make use of sterically crowded Lewis acids in order to introduce a steric repulsion between the nucleophile and the Lewis acid. As already reported, the polymerization of methacrylic esters with aluminum porphyrins proceeds via (porphyrinato)aluminum enolates 2 (Scheme 1), which are among the most sterically crowded enolates because of the presence of a bulky aluminum porphyrin counter species. Therefore, the aluminum enolate species 2 can coexist with a Lewis acid when the steric bulk around the Lewis acidic center is sufficiently large. An alternative to this approach focuses attention on the steric bulk of the nucleophile component 2 by using strategically designed aluminum porphyrins and methacrylates with a bulky ester group in order to utilize sterically fine-tuned initiators/growing species having sufficiently bulky substituents at the porphyrin periphery or in axial enolate species. This allows us to use sterically less-hindered, more common Lewis acids as accelerators. The third approach is to make use of Lewis acids with a sufficiently low ligand-exchange activity, where the steric protection of the initiator and Lewis acid is not necessary [20].

2.1.1
Polymerization of Methacrylic Esters via Enolatealuminum Porphyrins in the Presence of Methylaluminum Diphenolates and Dialcoholates [22,23]

Polymerization of methyl methacrylate (MMA) was initiated with methylaluminum tetraphenylporphyrin (1, X=Me) under irradiation with visible light. At a monomer conversion less than 10%, Lewis acids such as methylaluminum diphenolates 3a–i and methylaluminum dialcoholates 4a,b were added to the system in order to examine whether they could accelerate the polymerization (Scheme 2) (Table 1). For example, a CH_2Cl_2 solution (10 ml) of a mixture of MMA and 1 (X=Me) with a mole ratio of 217 was irradiated at 35 °C with xenon arc light ($\lambda > 420$ nm) to initiate the polymerization. During 2.5-h irradiation, all the molecules of the initiator 1 (X=Me) were transformed into the growing enolate species 2 (R=Me), where the conversionof the monomer was 6.1%, as estimated by ^{1}H NMR.

On the other hand, when 3 equiv of 3b with respect to 2 were added at room temperature to the above reaction mixture under diffuse light, a strikingly vig-

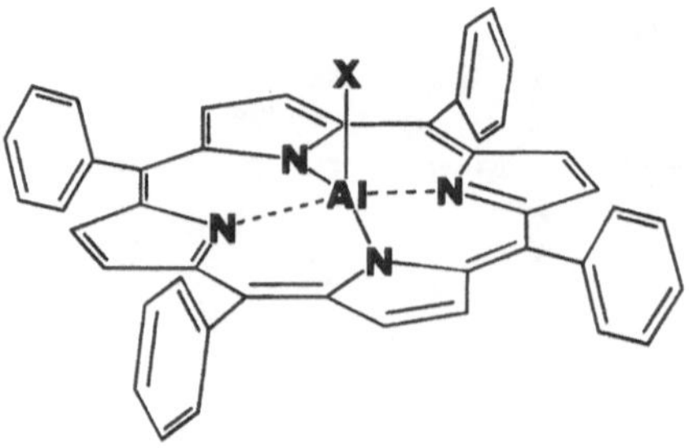

(TPP)AlX

Structure 1

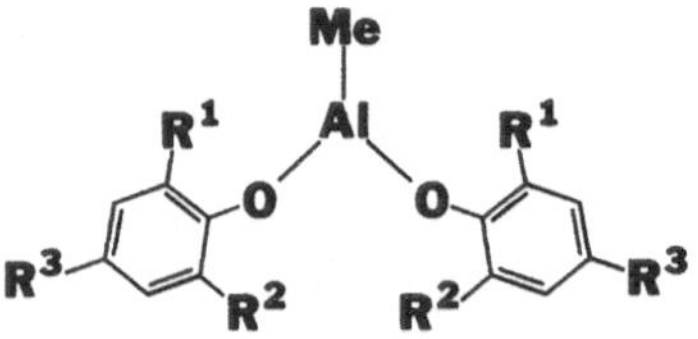

3a: $R^1 = {}^tBu$, $R^2 = R^3 = H$

3b: $R^1 = R^3 = {}^tBu$, $R^2 = H$

3c: $R^1 = {}^tBu$, $R^2 = H$, $R^3 = OMe$

3d: $R^1 = Ph$, $R^2 = R^3 = H$

3e: $R^1 = R^2 = {}^tBu$, $R^3 = Me$

3f: $R^1 = R^2 = R^3 = {}^tBu$

3g: $R^1 = R^2 = Cl$, $R^3 = H$

3h: $R^1 = R^2 = H$, $R^3 = {}^tBu$

3i: $R^1 = R^2 = H$, $R^3 = OMe$

3j: $R^1 = R^2 = Br$, $R^3 = H$

3k: $R^1 = R^2 = I$, $R^3 = H$

3l: $R^1 = R^2 = R^3 = Ph$

3m: $R^1 = R^2 = R^3 = OMe$

Structure 3

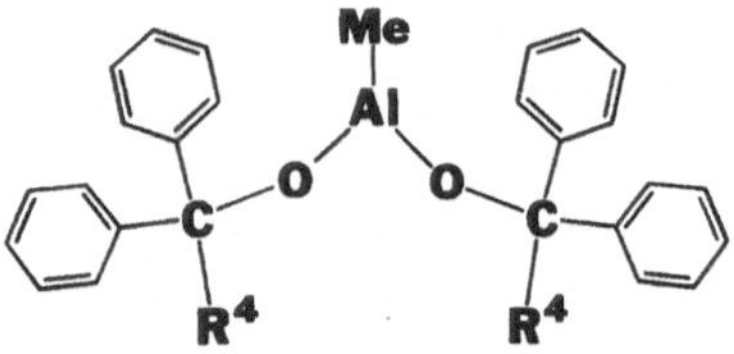

4a: $R^4 = H$

4b: $R^4 = Ph$

Structure 4

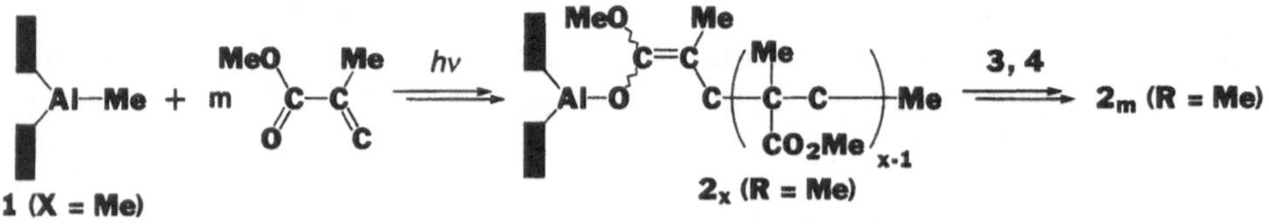

Scheme 2

Table 1. Polymerization of methyl methacrylate (MMA) via an enolatealuminum porphyrin (2, R=Me) in the presence of methylaluminum diolates $(MeAlX_2)$[a]

Run	$MeAlX_2$	$[MeAlX_2]_0/[2]_0$	Time/min	$Conv^b$/%	Mn^c	Mw/Mn^c
1	3a	3	0.5	100	21,500	1.10
2	3b	0.2	3	100	26,300	1.07
3	3b	2	10	100	23,700	1.07
4	3c	3	0.5	100	22,000	1.10
5	3d	3	0.75	92	27,700	1.42
6	3f	3	5	100	24,300	1.11
7	3g	0.5	0.5	74	31,800	1.16
8	3h	0.5	60	7	–	–
9	3i	0.5	60	8	–	–
10	4a	3	60	6	–	–
11	4b	3	60	91	24,000	1.04

[a] In CH_2Cl_2 under nitrogen, $[MMA]_0/[1\ (X=Me)]_0=200$, $[1,(X=Me)]_0=16.2$ mM, $[1,(X=Me)]_0=[2\ (R=Me)]_0$.
[b] Determined by ^1H NMR analysis of the reaction mixture.
[c] Estimated by GPC based on polystyrene standards.

Table 2. Polymerization of methyl methacrylate (MMA) via an enolatealuminum porphyrin (2, R=Me) in the presence of methylaluminum bis(2,4-di-*tert*-butylphenolate) (3b)[a]

Run	$[3b]_0/[2]_0$	Time/s	$Conv^b$/%	$Mn^c(Mn_{calc})$	Mw/Mn^c	Magnitude of acceleration[d]
1	0.2	5	21	5,500 (4300)	1.09	1400
2	1.0	5	62	14,400 (12,100)	1.10	14,000
3	3.0	3	100	25,500 (21,700)	1.07	46,200

[a] In CH_2Cl_2 under nitrogen, $[MMA]_0/[1\ (X=Me)]_0=200$, $[1,(X=Me)]_0=16.2$ mM, $[1,(X=Me)]_0=[2\ (R=Me)]_0$.
[b] Determined by ^1H NMR analysis of the reaction mixture.
[c] Estimated by GPC based on polystyrene standards.
[d] Mole ratio of MMA reacted per second before and after addition of 3b.

orous reaction took place with heat evolution to attain 100% conversion within only 3 s (Table 2, run 3). The number-average molecular weight of the polymer was 25,500, as estimated by GPC based on polystyrene standards, which was in good agreement with the expected value (21,700) by assuming that every molecule of 2 produced one polymer molecule. Of much interest is the cleanness of the reaction with the 2–3b system, as shown by the uniformity of the molecular weight of the produced polymer with the ratio of weight- and number-average molecular weights (*Mw/Mn*) being 1.09.

Similarly to 3b, among the methylaluminum diphenolates examined, 3a,c–g with substituents at the *ortho* positions of the phenolate ligands were quite effective for accelerating the polymerization, allowing the formation of narrow MWD PMMAs with the *Mn* values in fair agreement with those expected from the initial mole ratio of MMA to 1 (X=Me) (Table 1, runs 1–7). In terms of both the power of acceleration and the cleanness of polymerization, the diphenolates having *ortho tert*-butyl substituents (**3a–c, e,** and **f**) were the most favorable. On the other hand, when methylaluminum diphenolates having no *ortho* substituents (**3h** and **i**) were used, the polymerization was terminated just after they were added to the system (runs 8 and 9). As for methylaluminum dialcoholates (**4a, b**), the tertiary alcoholate **4b** gave a satisfactory result (run 11), whereas use of the secondary alcoholate **4a** resulted in termination of chain growth (run 10). When using **3a–g, 4b**, the dark reddish purple color of the polymerization system, characteristic of the growing enolate species **2** (R=Me), was observed to remain unchanged throughout the polymerization, while the polymerization systems immediately turned bright reddish purple characteristic of **1** (X=OR, OAr) upon addition of the Lewis acids **3h, i** or **4a**.

In the polymerization of MMA initiated with **1** (X=Me), the degree of acceleration became more pronounced as the amount of **3b** was increased (Table 2). If the added **3b** initiates the polymerization, the number of polymer molecules produced should increase proportionally to the amount of **3b**, and, consequently, the Mn value should decrease. However, irrespective of the ratio of **3b** to **2**, all the polymers formed were of narrow MWD, and the observed Mn values were always close to those expected from the mole ratios of the monomer reacted to **2**, indicating that the added **3** does not initiate but only accelerates the polymerization. In this regard, all the produced polymers in Table 2 were silent in GPC when monitored at 263 nm (2,4-di-*tert*-butylphenol: λ_{max} 263 nm, anisole: 269 nm), indicating no incorporation of the phenolate unit of **3b** into the polymer terminal. Furthermore, with **3b** alone, the polymerization of MMA did not take place at all under identical conditions.

The polymerization in the presence of bulky methylaluminum diphenolates proceeds with living character. Clear evidence for this was given by the sequential two-stage polymerization of MMA with the **2** (R=Me)–**3f** (1 equiv/1 equiv) system in CH_2Cl_2 at room temperature (Fig. 1), where 50 equiv of MMA with respect to **1** (X=Me) were polymerized up to 100% conversion (90 s) at the first stage, and after a 4-h interval at 25 °C, 200 equiv of MMA were charged to the mixture. In spite of a long interval between the first and second monomer feeds, the second stage polymerization ensued with heat evolution, and was complete within 60 s. The GPC profile showed a clear increase in the molecular weight of the polymer from 7000 [the first stage: Fig. 2 (I)] to 47,600 [the second stage: (II)], retaining the narrow MWD (Mw/Mn: from 1.12 to 1.05).

In addition to MMA, a variety of methacrylic esters were polymerized rapidly to the corresponding polymers with narrow MWDs in the presence of methylaluminum bis(2-*tert*-butyl-4-methoxyphenolate) (**3c**). The successful examples include ethyl methacrylate (EMA), isopropyl methacrylate ([iso]PMA), n-butyl methacrylate ([n]BMA), isobutyl methacrylate ([iso]BMA), benzyl methacrylate (BnMA), and dodecyl methacrylate (C_{12}MA), where the Mn values were all close to the predicted values (Mn_{calc}) with the Mw/Mn ratios below 1.1 (Table 3, runs 1–4, 6, 7). The polymerization of *tert*-butyl methacrylate ([t]BMA) is the only exception, where the monomer conversion hardly increased even after 24 h.

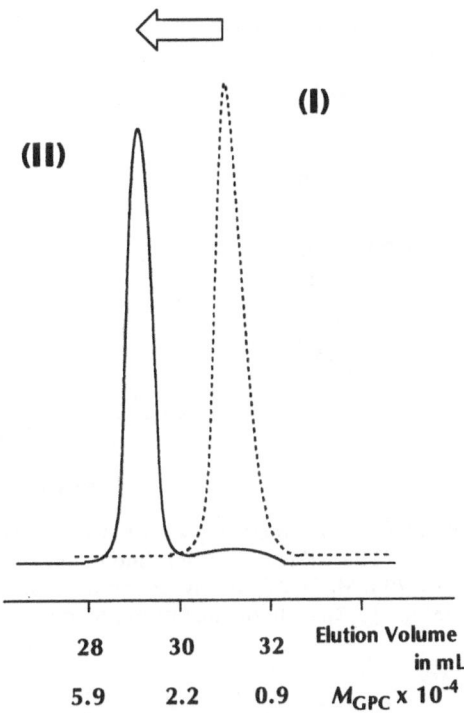

Fig. 1. Two-stage polymerization of methyl methacrylate (MMA) in CH_2Cl_2 at room temperature with the enolatealuminum porphyrin (2, R=Me)–methylaluminum bis(2,4,6-tri-*tert*-butylphenolate) (3f) (1.0:1.0) system. GPC profiles of the polymers formed at the first stage (*I*) $[MMA]_0/[1\ (X=Me)]_0=50$, 100% conversion; $Mn=7,000$, $Mw/Mn=1.12$ and the second stage (*II*) $[MMA]_0/[2]_0=200$, 100% conversion; $Mn=47,600$, $Mw/Mn=1.05$

Table 3. Polymerization of methacrylic esters via enolatealuminum porphyrins 2 in the presence of methylaluminum bis(2-*tert*-butyl-4-methoxyphenolate) (3c)[a]

Run	Monomer	Time/s	Conv[b]/%	Mn[c]	(Mn$_{calc}$)	Mw/Mn[c]
1	EMA	30	100	27,700	(23,000)	1.09
2	isoPMA	30	100	30,900	(25,600)	1.10
3	nBMA	30	100	34,900	(28,400)	1.07
4	isoBMA	30	100	36,300	(28,400)	1.07
5	tBMA	90	11	–	–	–
6	BnMA	90	100	32,900	(35,200)	1.08
7	C$_{12}$MA	90	64	30,600	(32,600)	1.09

[a]   In CH_2Cl_2 under nitrogen, $[monomer]_0/[1\ (X=Me)]_0/[3c]_0=200/1.0/0.5$, $[1,\ (X=Me)]_0=16.2$ mM.
[b]   Determined by ^1H NMR analysis of the reaction mixture.
[c]   Estimated by GPC based on polystyrene standards.

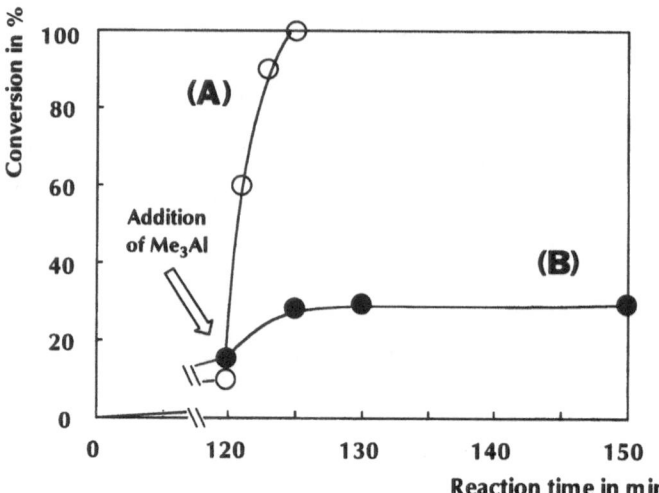

Fig. 2. Polymerizations of *tert*-butyl methacrylate (^{t}BMA) (*A*) and methyl methacrylate (MMA) (*B*) initiated with (TPP)AlMe (**1**, X=Me); [monomer]$_0$/[**1** (X=Me)]$_0$=100, [**1** (X= Me)]$_0$=17.8 (*A*) and 19.6 (*B*) mM, CH$_2$Cl$_2$ as solvent, room temperature. Effect of trimethylaluminum (Me$_3$Al) ([Me$_3$Al]$_0$/[**2**]$_0$=3.0) on the rate of polymerization

2.1.2
Polymerization of Methacrylic Esters via Enolatealuminum Porphyrins in the Presence of Trialkyl- and Triarylaluminum Compounds

Similar to methylaluminum diphenolates **3**, trialkylaluminums such as trimethylaluminum (Me$_3$Al), triisobutylaluminum (^{iso}Bu$_3$Al), and triphenylaluminum (Ph$_3$Al), by themselves, do not have the ability to initiate the polymerization of methacrylic esters under ordinary conditions. For investigating the acceleration effects of trialkylaluminums, the polymerization of *tert*-butyl methacrylate (^{t}B-MA) was first chosen, since it proceeds slowly even in the presence of methylaluminum diphenolates **3** as Lewis acids. However, the polymerization was dramatically accelerated when Me$_3$Al was added to the system. As exemplified in Fig. 2A, the polymerization of ^{t}BMA (100 equiv) with **1** (X=Me), which had proceeded to 7% conversion after 2-h irradiation at 35 °C, proceeded rapidly with vigorous heat evolution as soon as 3 equiv of Me$_3$Al with respect to **2** (R=^{t}Bu) were added at room temperature, and the complete consumption of ^{t}BMA was attained within only 5 min. The produced polymer exhibited a unimodal, sharp GPC peak, and the *M*n value increased linearly along the theoretical line (dotted, Fig. 3) calculated from the monomer-to-initiator (**1**, X=Me) mole ratio, keeping the *M*w/*M*n ratio in the range 1.05–1.06. It was also noted here that the addition of Me$_3$Al was accompanied by a rapid color change of the system from dark reddish purple due to **2** to bluish purple characteristic of methylaluminum tetra-

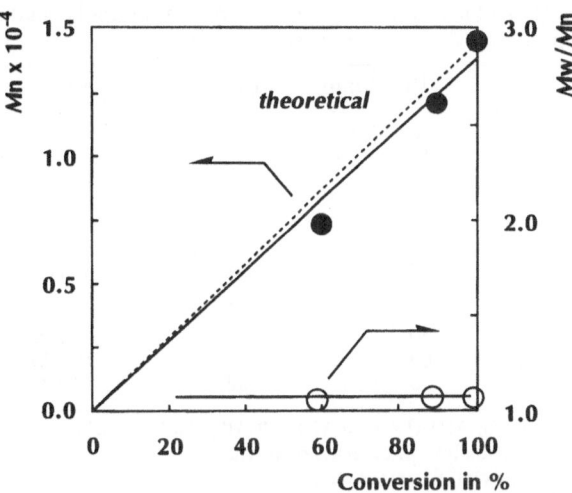

Fig. 3. Polymerization of *tert*-butyl methacrylate (^{t}BMA) initiated with (TPP)AlMe (**1**, X= Me) via an enolatealuminum porphyrin (**2**, R=^{t}Bu) in the presence of trimethylaluminum (Me$_3$Al); [^{t}BMA]$_0$/[**1** (X=Me)]$_0$/[Me$_3$Al]$_0$=100/1.0/3.0, [**1**]$_0$=17.8 mM, CH$_2$Cl$_2$ as solvent, rt. Relationship between *Mn* (●) [*Mw*/*Mn* (○)] of the polymer and conversion

phenylporphyrin (**1**, X=Me), suggesting that the polymerization proceeded via dimethylaluminum enolate as the active species.

In contrast, when the monomer was methyl methacrylate (MMA), the polymerization was terminated upon addition of Me$_3$Al under similar conditions. For example, when 3 equiv of Me$_3$Al with respect to **2** (R=Me) were added to the system ([MMA]$_0$/[**1** (X=Me)]$_0$=100, 9% conversion), heat evolution, although observed at the very early stage, subsided within only 1–2 min. At this stage, the color of the solution had already turned bluish purple. The monomer conversion after 5 min was only 30%, and it no longer increased upon prolonged reaction for 2 h (Fig. 2B). Use of bulkier ^{iso}Bu$_3$Al in place of Me$_3$Al for the polymerization of MMA ([MMA]$_0$/[**2** (R=Me)]$_0$/[^{iso}Bu$_3$Al]$_0$=200/1.0/3.0) gave an essentially similar result, although the monomer conversion finally attained was higher (64%) (run 2, Table 4). The *Mn* of the polymer formed at this conversion (20,200) was much higher than that expected from the ratio of MMA reacted to **1** (X=Me) (12,800), and the MWD was broad (*Mw*/*Mn*=1.41). On the contrary, when Ph$_3$Al [3 equiv with respect to **2** (R=Me)] was added to the system at room temperature ([MMA]$_0$/[**1**]$_0$=200, 3% conversion), a fairly rapid polymerization took place with heat evolution, and the monomer was completely consumed within 90 min (run 3, Table 4). The *Mn* of the produced polymer (22,300) was close the expected value of 20,000 and the MWD was satisfactorily narrow (*Mw*/*Mn*=1.18). In this case, the system retained the original color characteristic of **2** (R=Me) throughout the polymerization.

Table 4. Polymerization of methyl methacrylate (MMA) via an enolatealuminum porphyrin (2, R=Me) in the presence of trialkylaluminums (R_3Al)[a]

Run	R_3Al	Time[d]/min	Conv[b]/%	Mn[c]	Mw/Mn[c]
1	Me_3Al	0	4	–	–
		10	9	–	–
		30	9	–	–
2	$^{iso}Bu_3Al$	0	6	–	–
		10	63	19,500	1.37
		30	64	20,200	1.41
3	Ph_3Al	0	3	–	–
		30	46	8,600	1.19
		60	79	16,700	1.17
		90	100	22,300	1.18

[a] In CH_2Cl_2 under nitrogen, $[MMA]_0/[1\ (X=Me)]_0/[R_3Al]_0=200/1.0/3.0$, $[1, (X=Me)]_0=16.2$ mM.
[b] Determined by ^1H NMR analysis of the reaction mixture.
[c] Estimated by GPC based on polystyrene standards.
[d] After addition of R_3Al.

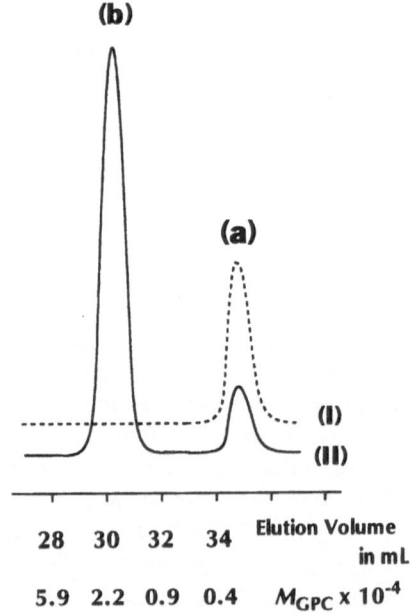

Fig. 4. Polymerizations of *tert*-butyl methacrylate (^tBMA) initiated with (TPP)AlMe (1, X= Me); $[^tBMA]_0/[1\ (X=Me)]_0=100$, $[1]_0=17.8$ mM, CH_2Cl_2 as solvent. GPC profiles of the polymers formed *I* After 90-h irradiation at 35 °C, 25% conversion, and *II* In 60 min after addition of trimethylaluminum (Me_3Al), $([Me_3Al]_0/[2\ (R=^tBu)]_0=0.25)$ at rt, 100% conversion

At a lower temperature, in contrast, even Me_3Al was able to accelerate the polymerization of MMA. At $-40\ °C$, for example, after a 3 molar amount of Me_3Al had been added to the polymerization system ($[MMA]_0/[2\ (X=Me)]_0=100$, 11% conversion) 91% conversion was attained in the following 1 h. Furthermore, the polymer formed here had a narrow MWD as indicated by the Mw/Mn ratio of 1.17, and the Mn value (9700) was close to that expected (9100) from the mole ratio of the monomer reacted to the initiator. In this case, the polymerization was observed to proceed without any change in color of the system. Thus, the polymerization temperature is one of the important factors to achieve the clean and rapid polymerization with trialkylaluminums.

In this regard, when 0.25 equiv of Me_3Al with respect to the growing species **2** was added at room temperature to the polymerization system of *tert*-butyl methacrylate (^tBMA) ($[^tBMA]_0/[1\ (X=Me)]_0=100$) at 25% conversion, a part of the prepolymer [peak a in Fig. 4 (I)] remained unreacted even after complete monomer consumption had been attained [Fig. 4 (II)]. When the amount of Me_3Al relative to **2** was increased from 0.25 to 0.5, peak a became much smaller, while the molecular weight corresponding to the major peak (peak b) decreased. On the other hand, when the ratio $[Me_3Al]_0/[2]_0$ exceeded unity (1.5 and 3.0), peak a disappeared completely to give a unimodal MWD product. At $[Me_3Al]_0/[2]_0\geq1$, the Mn value of the final product was no longer affected by the ratio $[Me_3Al]_0/[2]_0$ (Fig. 5). Thus, the number of polymer molecules vs. **2**

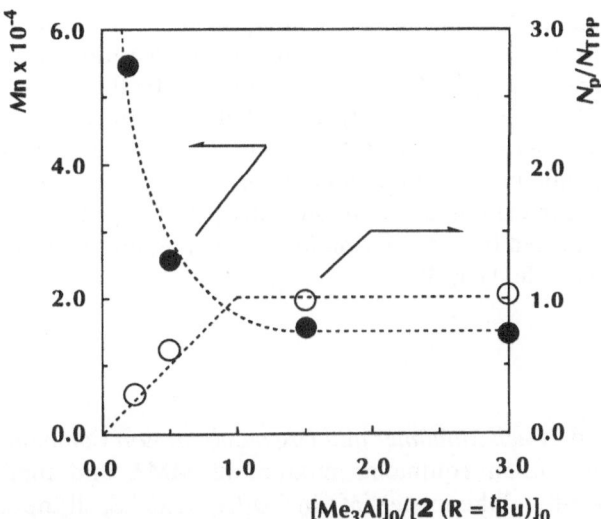

Fig. 5. Polymerization of *tert*-butyl methacrylate (^tBMA) initiated with (TPP)AlMe (**1**, X= Me) via an enolatealuminum porphyrin (**2**, R=^tBu) in the presence of trimethylaluminum (Me_3Al); $[^tBMA]_0/[1\ (X=Me)]_0=100$, $[1]_0=17.8$ mM, CH_2Cl_2 as solvent, rt. Correlations of the number-average molecular weight (Mn) and the ratio of the numbers of polymer molecules to **2** (N_p/N_{TPP}) with the initial mole ratio of Me_3Al to **2**

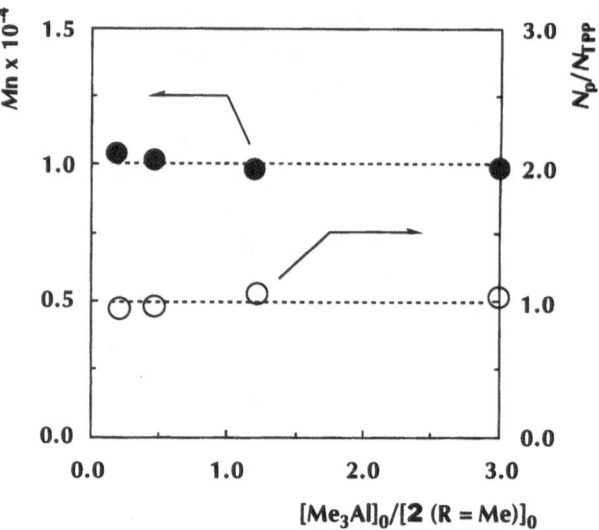

Fig. 6. Polymerization of methyl methacrylate (MMA) initiated with (TPP)AlMe (1, X=Me) via an enolatealuminum porphyrin (2, R=Me) in the presence of trimethylaluminum (Me$_3$Al; [MMA]$_0$/[1 (X=Me)]$_0$=100, [1]$_0$=19.6 mM, CH$_2$Cl$_2$ as solvent, –40 °C. Correlations of the number-average molecular weight (Mn) and the ratio of the numbers of polymer molecules to 2 (N$_p$/N$_{TPP}$) with the initial mole ratio of Me$_3$Al to 2

(N$_p$/N$_{TPP}$), calculated from the Mn of the polymer corresponding to peak b [Fig. 4 (II)] and the initial monomer-to-initiator mole ratio, was increased with increasing the ratio [Me$_3$Al]$_0$/[2]$_0$ up to unity, but no further increase in N$_p$/N$_{TPP}$ resulted after the ratio [Me$_3$Al]$_0$/[2]$_0$ exceeded unity. In sharp contrast, in the case of the polymerization of MMA at –40 °C with varying mole ratios of [Me$_3$Al]$_0$/[2]$_0$, only the rate of polymerization was changed, but in any case the produced polymer exhibited a unimodal sharp GPC peak with the Mn value close to that expected from the formation of one polymer molecule from every molecule of 2 (R=Me) (Fig. 6).

2.1.3
NMR Studies

(I) Interaction Between Monomer and Organoaluminum Compounds. In the ^{1}H NMR spectrum of an equimolar mixture of MMA and methylaluminum bis(2,4,6-tri-*tert*-butylphenolate) (3f) in CD$_2$Cl$_2$ at 25 °C, all the signals due to MMA clearly shifted downfield from those of MMA alone, where the signals due to CH$_2$ (δ 6.00 and 5.49), CH$_3$O (δ 3.66), and CH$_3$ (δ 1.86) of MMA appeared at δ 6.41 and 5.81, 3.92, and 1.98, respectively. In the ^{13}C NMR spectrum, the signals due to, e.g. C=O (a, δ 168.3), CH$_2$ (c, δ 125.6), and CH$_3$O (d, 52.2) of MMA also shifted downfield to δ 173.3 (a'), 133.8 (c'), and 56.5 (d'), respectively, in the pres-

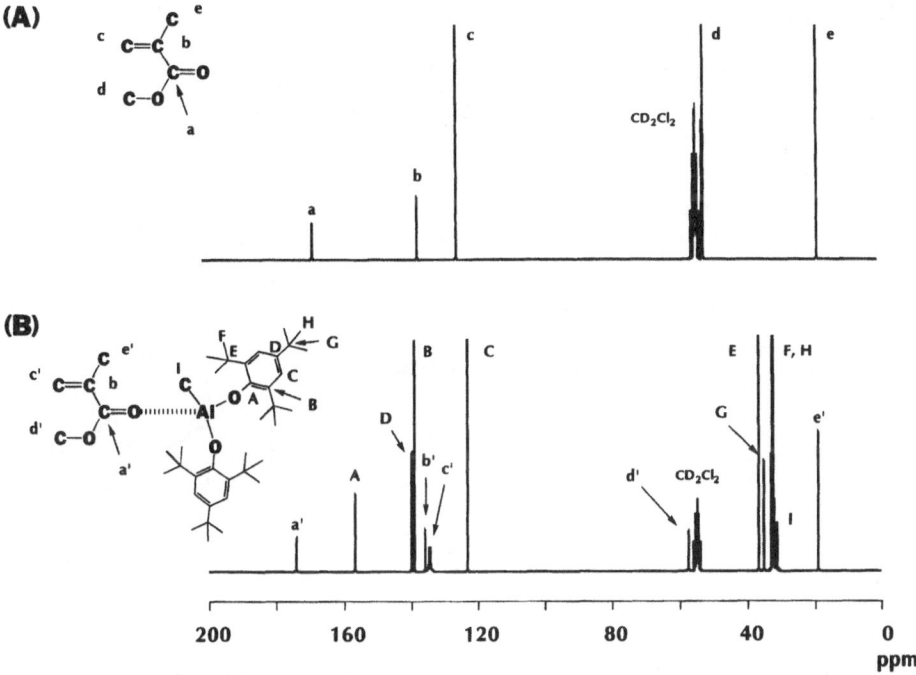

Fig. 7. ^{13}C NMR spectra in CD$_2$Cl$_2$ at 25 °C of *A* Methyl methacrylate (MMA) and *B* An equimolar mixture of MMA and methylaluminum bis(2,4,6-*tert*-butylphenolate) (3f)

ence of an equimolar amount of **3f** (Fig. 7B). As for the signals of **3f** [21], the *C*-O signal of the phenolate ligands most clearly shifted from δ 152.6 to 155.7 (A). These observations demonstrate the occurrence of a coordinative interaction between the carbonyl group of MMA and **3f**.

In the ^{13}C NMR spectrum of an equimolar mixture of MMA and Me$_3$Al in CD$_2$Cl$_2$ at 25 °C, a downfield shift was again observed for the *C*=O signal of MMA from δ 168.3 to δ 172.2. On the other hand, for an equimolar mixture of MMA and Ph$_3$Al, the downfield shift of the MMA *C*=O signal was very small (0.1 ppm) compared with those for the MMA–**3f** and MMA–Me$_3$Al systems under similar conditions.

(II) Ligand-Exchange Reaction Between the Growing Species and Organoaluminum Compounds. A C$_6$D$_6$ solution of an equimolar mixture of the living polymer of MMA (**2**, R=Me; [MMA]$_0$/[**1** (X=Me)]$_0$=10, 100% conversion) and **3f** was studied by ^{1}H NMR. If the ligand-exchange reaction takes place between **2** (R= Me) and **3f**, **1** (X=Me) and/or **6** (R^1=R^2=R^3=^{t}Bu) should be formed (Scheme 3), which are easily detectable by the characteristic signals at δ –5.79 and –0.26 due, respectively, to the Me-Al group of **1** (X=Me) and the ^{t}Bu group of **6** (R^1=R^2=R^3= ^{t}Bu). However, throughout the observation over a period of 24 h at 25 °C, neither of these two signals was detected.

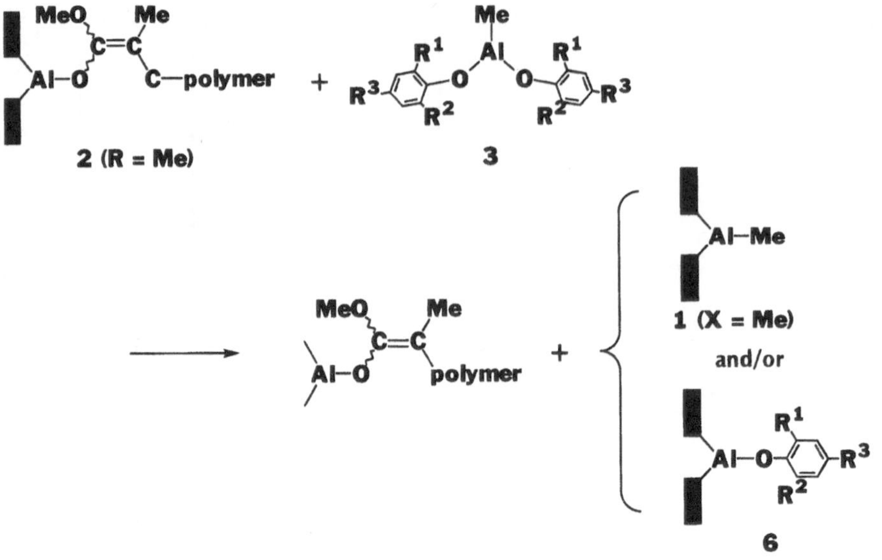

Scheme 3

In sharp contrast, when Me$_3$Al was mixed with **2** (R=^{t}Bu; [^{t}BMA]$_0$/[1 (X= Me)]$_0$=5, 100% conversion) under similar conditions, the color of the solution immediately turned from dark reddish purple to bluish purple. The ^{1}H NMR spectrum of this mixture in C$_6$D$_6$ showed the appearance of a new signal due to **1** (X=Me) (f, δ −5.79) with an intensity ratio of 3:8 to that of the pyrrole-β protons (8H) of the porphyrin ligand, while a characteristic signal at δ −0.33 ppm due to the enolate ^{t}Bu group of **2** (R=^{t}Bu) (a) [22] had completely disappeared (Fig. 8). This observation indicates the occurrence of a rapid ligand exchange between **2** (R=^{t}Bu) and Me$_3$Al (Scheme 4). The ligand-exchange reaction also occurred in the **2** (R=Me)–Me$_3$Al system.

Figure 9A shows the ^{1}H NMR spectrum in CD$_2$Cl$_2$ at 25 °C of **2** (PMMA-d_8), prepared from MMA-d_8 and **1** (X=Me) with the ratio of 10:1, where only the terminal methyl group originating from **1** (X=CH$_3$) (δ 0.70) is observed besides the signals due to the porphyrin ligand of the aluminum enolate species [a, δ 9.10 (pyrrole-β-H); b, δ 8.21 (Ph-o-H); c, δ 7.80 (Ph-m, p-H)]. Upon addition of 3 equiv of Me$_3$Al to this enolate solution at 25 °C, the spectrum changed (Fig. 9B), where a new signal (f') due to the Me-Al group of **1** (X=Me) appeared in addition to some new signals (e', g') assignable to the Me-Al groups of the dimethylaluminum enolate and excess Me$_3$Al. Moreover, the signals due to the porphyrin ligand had shifted slightly but definitely to those for **1** (X=Me) [a', δ 9.05 (pyrrole-β-H); b', δ 8.19 (Ph-o-H); c', δ 7.77 (Ph-m, p-H)], although the signal due to the terminal methyl group shifted only little from d to d', as expected.

Contrary to the above results at 25 °C, the ligand-exchange reaction (Scheme 4) was found to be greatly retarded by lowering the temperature.

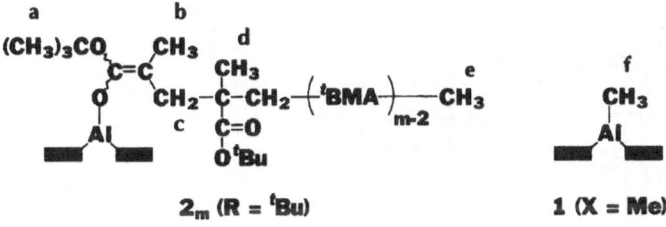

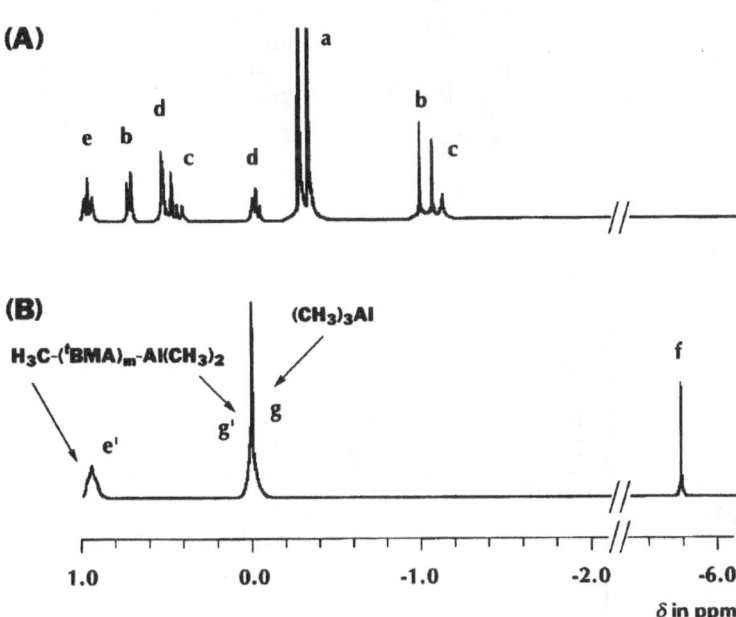

Fig. 8. ^{1}H NMR spectra in CD$_2$Cl$_2$ at 25 °C of *A* The enolatealuminum porphyrin (2, R=^{t}Bu; [^{t}BMA]$_0$/[1 (X=Me)]$_0$=5, 100% conversion) and *B* The reaction mixture of 2 (R=^{t}Bu) and trimethylaluminum (Me$_3$Al); [Me$_3$Al]$_0$/[2]$_0$=3.0, [2]$_0$=25.0 mM

Scheme 4

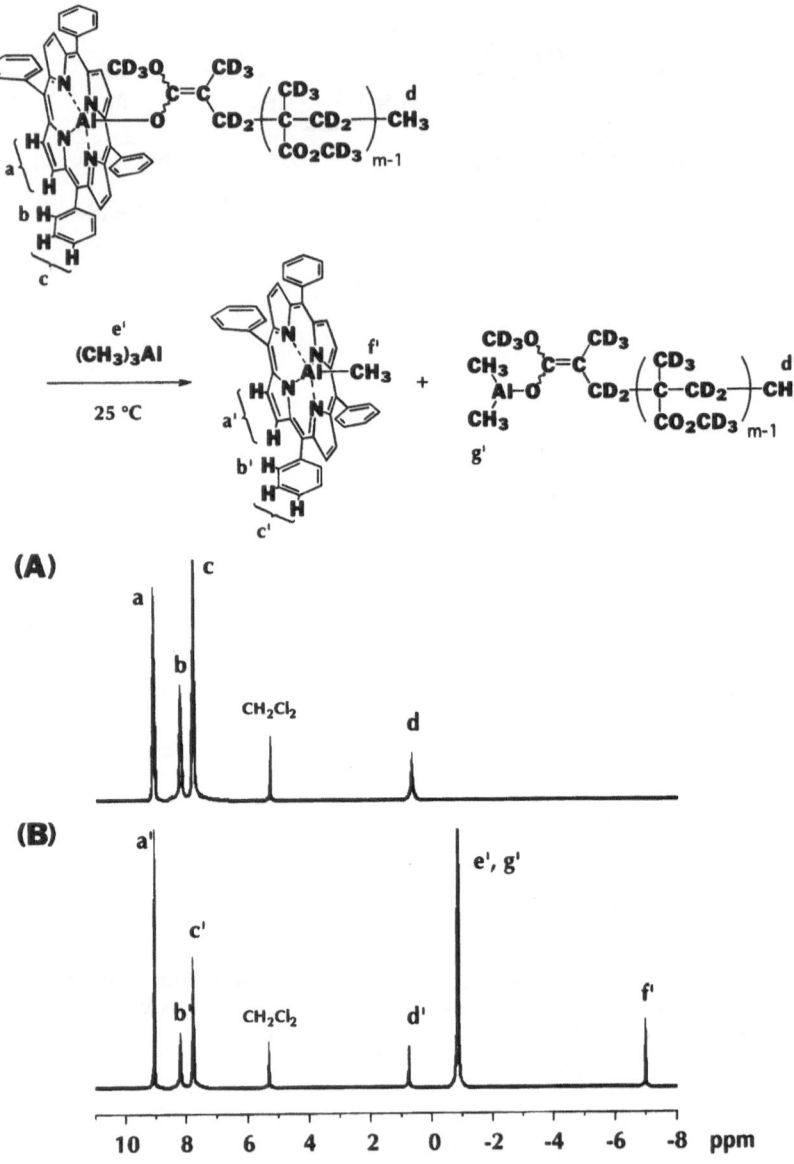

Fig. 9. ^{1}H NMR spectra in CD$_2$Cl$_2$ at 25 °C of A The enolatealuminum porphyrin (2, PMMA-d_8; [MMA-d_8]$_0$/[1 (X=Me)]$_0$=10, 100% conversion) and B The reaction mixture of 2 (PMMA-d_8) and trimethylaluminum (Me$_3$Al); [Me$_3$Al]$_0$/[2]$_0$=3.0, [2]$_0$=25.0 mM

Figure 10 shows the change in the mole fraction of 1 (X=Me): [1 (X=Me)]/([1 (X=Me)]+[2 (PMMA-d_8)]), as determined from the relative intensity of the signals f' to a and a' (Fig. 9), after addition of 3 equiv of Me$_3$Al to 2 (PMMA-d_8) at

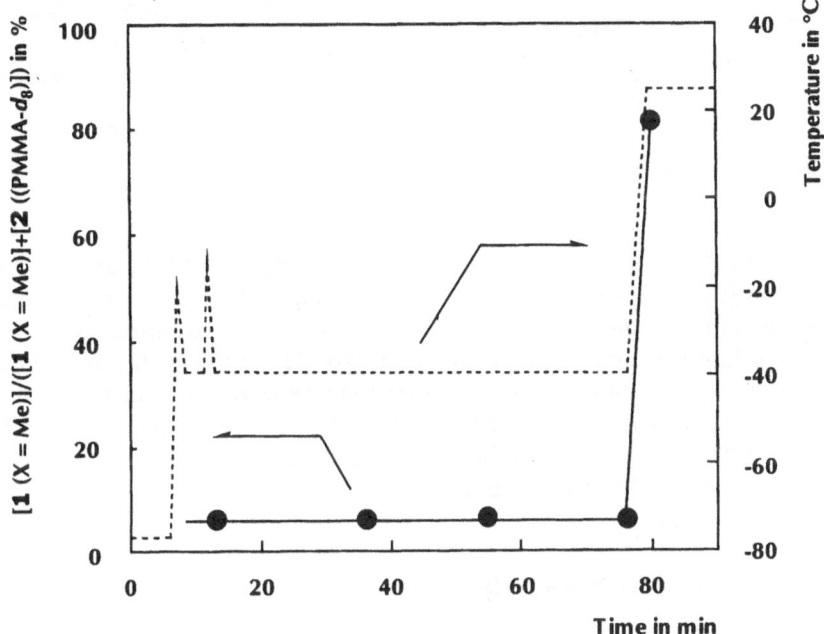

Fig. 10. Time course of the reaction of the enolatealuminum porphyrin (2, PMMA-d_8; [MMA-d_8]$_0$/[1 (X=Me)]$_0$=10, 100% conversion) with trimethylaluminum (Me$_3$Al) in CD$_2$Cl$_2$ at –40 to 25 °C; [Me$_3$Al]$_0$/[2]$_0$=3.0, [2]$_0$=25.0 mM, in NMR tube (Φ=5 mm). [1 (X= Me)]/([1 (X=Me)]+[2 (PMMA-d_8)]): Determined from the relative intensity of the ^{1}H NMR signals at δ –6.97 (1, Al-CH$_3$) to δ 9.0–9.1 (f' to a and a' in Fig. 10)

–40 °C. Although a trace amount of 1 (X=Me) was accidentally generated at the initial stage until the probe temperature had settled, its mole fraction was never increased throughout the observation over a period of 75 min. On the other hand, when the probe temperature was jumped to 25 °C, a steep increase in the mole fraction of 1 (X=Me) was observed.

2.1.4
Polymerizations of Methacrylic Esters Initiated with Diethylaluminum Enolate

In connection with the ligand-exchange reaction between the enolatealuminum porphyrin 2 and Me$_3$Al (Scheme 4), the diethylaluminum enolate 5, a model compound of the exchanged product, was synthesized according to Scheme 5, and tested as initiator for the polymerizations of methyl methacrylate (MMA) and *tert*-butyl methacrylate (^{t}BMA). For example, when 100 equiv of MMA were added to a CH$_2$Cl$_2$ solution of 5 at room temperature, the polymerization was actually initiated by heat evolution, but terminated only at 10.3% conversion, affording a low molecular-weight PMMA possessing a very broad MWD as indi-

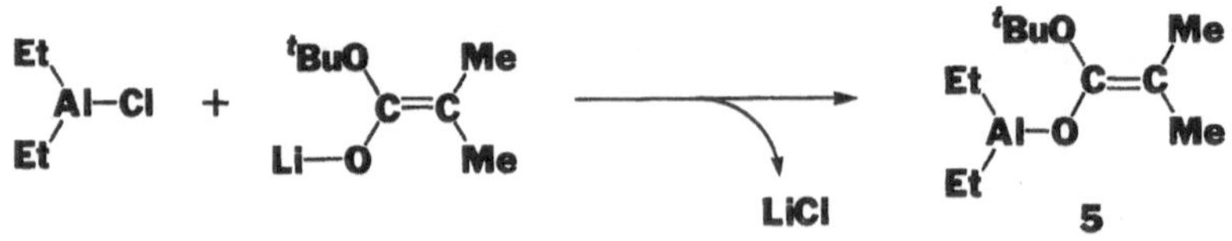

Scheme 5

cated by the Mw/Mn ratio of 5.6. On the other hand, as for ^{t}BMA under similar conditions ($[^t$BMA$]_0/[5]_0$=100), the enolate 5 brought about a very rapid polymerization, and the complete monomer consumption was attained within minutes, affording the narrow MWD polymer (Mw/Mn=1.05) with Mn of 34,600. The observed Mn value, although higher than expected, was observed to increase almost linearly to 77,100 and 102,000 retaining the narrow MWD (Mw/Mn 1.05–1.13) [23], when the initial monomer-to-initiator mole ratio was increased to 200 and 300, respectively.

2.1.5
Discussion on the Mechanistic Aspects

Methylaluminum *ortho*-substituted phenolates 3a–g, methylaluminum bis(triphenylmethanolate) (4b), and Ph$_3$Al as Lewis acid additives can dramatically accelerate, even at room temperature, the polymerization of MMA via 2 without damaging the living character of polymerization. On the other hand, when methylaluminum phenolates bearing no *ortho*-substituents 3h,i, methylaluminum bis(diphenylmethanolate) (4a), Me$_3$Al, and ^{iso}Bu$_3$Al are used under similar conditions, the polymerization is terminated. When the temperature is lowered to, e.g. –40 °C, even Me$_3$Al can be used to accelerate the polymerization of MMA without any loss of the living character. Together with the results of the NMR investigations, the accelerated living polymerization of MMA with the 2–Lewis acids (3a–g) systems is concluded to be the result of (1) coordinative activation of MMA for the nucleophilic attack of 2 (Fig. 7) and (2) suppression of the undesired degradative attack of 2 to the Lewis acidic center (Scheme 3) because of the steric repulsion between the bulky porphyrin ligand of 2 and the substituents at the *ortho* positions of the phenolate ligands of 3a–g. The same may be true for the 2–Ph$_3$Al system.

When the substituents of the Lewis acid are smaller, as typically in the case of Me$_3$Al, the undesired ligand-exchange reaction (Scheme 4) competitively occurs, leading to the formation of dimethylaluminum enolate 7 at the expense of change of 2 into methylaluminum porphyrin (1, X=Me) (Fig. 9). As already reported, 1 (X=Me) has no ability to initiate the polymerization of methacrylic esters without visible light irradiation. Furthermore, dimethylaluminum enolate is not a suitable growing species for the living polymerization of MMA, taking into account the fact that the polymerization initiated with the diethylaluminum enolate 5 at room temperature is terminated at a very early stage. Therefore, the ligand-exchange reaction (Scheme 4) leads to the termination of chain growth in the polymerization of MMA.

Similar to the above, the ligand-exchange reaction (Scheme 4) actually occurs at room temperature when the monomer is *tert*-butyl methacrylate (^{t}BMA) (Fig. 8), but the resulting dimethylaluminum enolate 7 (R=^{t}Bu) can initiate the rapid polymerization of ^{t}BMA in a fairly controlled fashion, considering the quantitative formation of a narrow MWD P^{t}BMA using 5 as the initiator. Therefore, when Me$_3$Al is added to the living P^{t}BMA (2, R=^{t}Bu) at a ratio below unity, only a part of the polymer molecules are transformed into the more active form 7 (R=^{t}Bu), and the formation of a bimodal MWD P^{t}BMA results (Fig. 4). This is also supported by the results shown in Fig. 5. At a lower temperature, such as –40 °C, the ligand-exchange reaction (Scheme 4) is extremely reluctant to occur (Fig. 10), so that the polymerization is accelerated by Me$_3$Al without losing the living character (Fig. 6), similar to the case using the 2–bulky 3 systems.

2.1.6
Polymerizations of Methacrylic Esters Initiated with Sterically Crowded Growing Species–Less Bulky Organoaluminum Compound Systems [24]

Of key importance in the Lewis acid promoted living anionic polymerization of methacrylic esters with aluminum porphyrin is how to suppress the undesired reaction between the nucleophile (2_m) and the Lewis acid, leading to termination of polymerization (Fig. 11). As mentioned in previous sections, one of our approaches was to make use of sterically crowded Lewis acids such as methylaluminum bis(*ortho*-substituted phenolates). This section focuses attention on the steric bulk of the nucleophile component (2_m), by using strategically designed aluminum porphyrins and some other methacrylates, for the purpose of understanding the scope and limitation of this method (Fig. 12).

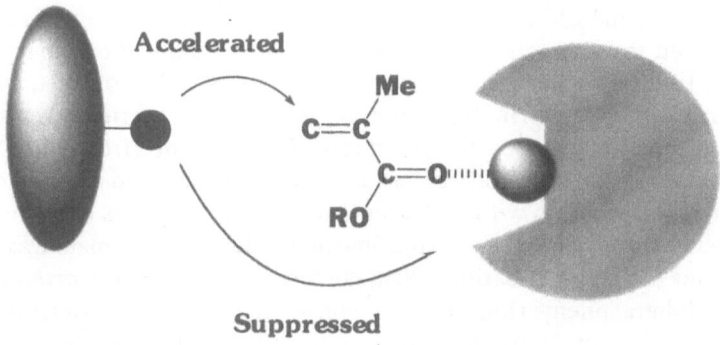

Fig. 11. The basic concept of the Lewis acid assisted "high-speed living anionic polymerization" of methacrylic esters

(A)   (B)

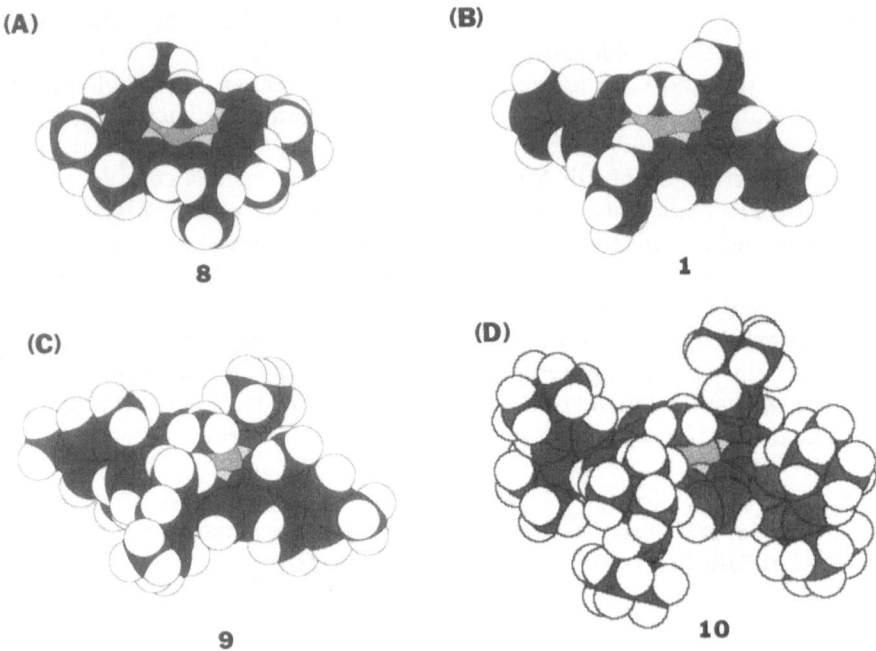

8   1

(C)   (D)

9   10

Fig. 12. Space-filling representations of methylaluminum porphyrins (1, 8–4)

As already mentioned above, the polymerization of MMA initiated with (TPP)AlMe (**1**, X=Me; Fig. 12B) was accelerated during the early stage after the addition of $^{iso}Bu_3Al$, but soon terminated, since $^{iso}Bu_3Al$ reacted rather easily with the growing enolate species (**2**, R=Me) (Table 5, run 2). When a sterically less-crowded initiator, such as methylaluminum etioporphyrin I (**8**, Fig. 12A), was used under otherwise identical conditions, the polymerization stopped immediately after the addition of $^{iso}Bu_3Al$ (Table 5, run 1). Thus, in these two cases, an undesired reaction took place between the nucleophilic growing species and $^{iso}Bu_3Al$, leading to the termination of chain growth. In contrast, when methyl(tetramesitylporphyrinato)aluminum (**9**, Fig. 12C), a sterically more-crowded initiator than **2**, was used, the polymerization proceeded from **8** to 61% monomer conversion in 10 min after the addition of $^{iso}Bu_3Al$, and reached complete monomer consumption within 30 min (Table 5, run 3). In this case, the color of the system characteristic of the (porphyrinato)aluminum enolate was retained throughout the polymerization. Thus, the methyl groups at the *ortho* positions of the peripheral phenyl rings in **9** were considered to serve as a steric barrier for the access of $^{iso}Bu_3Al$ to the nucleophilic center. Nevertheless, the barrier seemed not to be sufficient when less bulky Me_3Al and Et_3Al were used as Lewis acids. In connection with this observation, when methylaluminum tetraphenylporphyrin carrying *tert*-butyl groups at the *meta* positions (**10**, Fig. 12D) was used in place of **9**, polymerization in the presence of $^{iso}Bu_3Al$ was terminated at

8 : R^1 = H, R^2 = Me, R^3 = Et

9 : R^1 = (2',4',6'-Me$_3$)Ph, R^2 = R^3 = H

10 : R^1 = (3',5'-^{t}Bu$_2$)Ph, R^2 = R^3 = H

(1 : R^1 = Ph, R^2 = R^3 = H)

Structure 8-10

Table 5. Polymerization of methacrylic esters via enolatealuminum porphyrins 2 in the presence of methylaluminum bis(2-*tert*-butyl-4-methoxyphenolate) (3c)[a]

Run	Initiator	R'$_3$Al	Monomer	Time[b]/ min	Conv.[c]/ %	Mn[d]	(Mn$_{calc}$)[e]	Mw/Mn[d]
1	8	iso-Bu$_3$Al	MMA	0	9	–		–
				10	11	–		–
				30	12	–		–
2	1	iso-Bu$_3$Al	MMA	0	6	–		–
				10	63	–		–
				30	64	20,200	(12,800)	1.41
4	9	iso-Bu$_3$Al	MMA	0	8	–		–
				10	61	–		–
				30	100	21,400	(20,000)	1.06
5	10	iso-Bu$_3$Al	MMA	0	7	–		–
				10	53	–		–
				30	53	11,100	(10,600)	1.35
6	1	iso-Bu$_3$Al	EMA	0	5	–		–
				10	100	26,500	(22,800)	1.19
7	1	iso-Bu$_3$Al	PMA	0	12	–		–
				10	100	27,400	(25,600)	1.18

[a] In CH$_2$Cl$_2$ under nitrogen, [initiator]$_0$/[monomer]$_0$/[R'$_3$Al]$_0$=1/200/3, [initiator]$_0$=16.2 mM.
[b] After addition of R'$_3$Al. Determined by ^{1}H NMR analysis of the reaction mixture.
[c] Estimated by GPC based on polystyrene standards. [d]By GPC based on polystyrene standards.
[e]Mn$_{calc}$=molecular weight of monomer×([monomer]$_0$/[initiator]$_0$)×(conversion/100).

53% monomer conversion, giving a polymer with a broader MWD (Table 5, run 4). Thus, even bulky *tert*-butyl groups, when introduced at the *meta* positions of the phenyl rings of 10, are not able to form an effective barrier to suppress any undesired reaction.

The enolate species 2, derived from methacrylates with bulkier ester groups than MMA, are sterically protected against the access of $^{iso}Bu_3Al$ under the above-mentioned conditions, even when the porphyrin moiety is a non-*ortho*-substituted tetraphenylporphyrin. An example is shown by the polymerization of ethyl methacrylate (EMA) using 1 (X=Me) as an initiator, where the growing species have an EtO group in the terminal enolate unit 2 (R=Et). After the addition of $^{iso}Bu_3Al$ to the system, polymerization proceeded to 100% monomer conversion within 10 min. The Mn of the produced polymer was close to the expected value, and the MWD was narrow (Table 5, run 5). A similar result was obtained for the polymerization of isopropyl methacrylate (PMA) with the 1 (X=Me)–$^{iso}Bu_3Al$ system, which quantitatively gave a narrow MWD poly(methacrylate) with a predicted Mn (Table 5, run 6).

From these results, not only the steric bulk of the Lewis acid (monomer activator), but also that of the nucleophilic growing species 2, is important for realizing the Lewis acid assisted, controlled anionic polymerization; the basic concept involving a sterically separated nucleophile–electrophile model is thus clearly demonstrated.

2.2
Living Polymerization of Methacrylic Ester with Aluminum Porphyrin–Organoboron Compound Systems [19]

2.2.1
Polymerization of Methacrylic Esters via Enolatealuminum Porphyrins in the Presence of Organoboron Compounds

Since organoboron compounds are much more reluctant than organoaluminum compounds to undergo a ligand-exchange reaction with nucleophiles, they could be used as Lewis acid accelerators without any steric protection of the nucleophilic and Lewis acidic centers [25]. Various boron compounds were tested as Lewis acid accelerators for the polymerization of MMA initiated with methylaluminum porphyrin (1, X=Me). First examined were $BF_3 \cdot OEt_2$ and BCl_3, which are highly acidic and commonly used for organic synthesis. For example, a C_6H_6 solution (10 ml) of a mixture of MMA and 1 (X=Me) with a mole ratio of 200 was irradiated at 35 °C with xenon arc light ($\lambda > 420$ nm) to initiate the polymerization. During 3-h irradiation, all the molecules of the initiator (1, X=Me) were transformed into the growing enolate species (2, R=Me), and the conversion of MMA reached 6.2%, as determined by ^1H NMR [26]. At this stage, the irradiation was stopped and an equimolar amount of $BF_3 \cdot OEt_2$ with respect to 2 (R=Me) was added to the system at room temperature, whereupon the color of the solution immediately turned from dark reddish brown characteristic of 2 (R=Me) to dark green, indicating the formation of (TPP)AlF (1, X=F) [27]. After 10 min under diffuse light, the monomer conversion was again measured, but it was virtually unchanged from the beginning (6.9%), and no longer increased upon prolonged reaction for 20 h under the same conditions. Use of BCl_3 ($[BCl_3]_0/[2 \ (R=Me)]_0 = 1.0$) in place of $BF_3 \cdot OEt_2$ gave a similar result, where the color of the solution was changed to dark reddish purple characteristic of 1 (X=Cl) as soon as BCl_3 was added. When triphenyl borate or trimethylboroxin was used as the Lewis acid under similar conditions, the polymerization was again terminated, where the color of the system immediately turned to bright reddish purple as typically observed for 1 (X=OR, OAr).

In marked contrast to the above cases, an organoboron compound such as triphenylboron $[(C_6H_5)_3B]$ did accelerate the polymerization without termination. Also in contrast, the polymerization system retained the original color characteristic of 2 (R=Me) throughout the polymerization. For example, when $(C_6H_5)_3B$ was added at room temperature to the polymerization mixture of MMA with 1 (R=Me) at a mole ratio of $(C_6H_5)_3B$ to 1 (R=Me) of 1.0, a modest heat evolution occured, and the polymerization was about 12-times faster (Fig. 13, ●). Although the observed acceleration effect was lower than that of, e.g. methylaluminum bis(*ortho*-substituted phenolates) as Lewis acids (3a–g), it was almost comparable to that of $(C_6H_5)_3Al$ [28]. When the mole ratio of $(C_6H_5)_3B$ to the growing species 2 was doubled under similar conditions ($[MMA]_0/[1 \ (X=Me)]_0=300$), the polymerization was more explicitly accelerated,

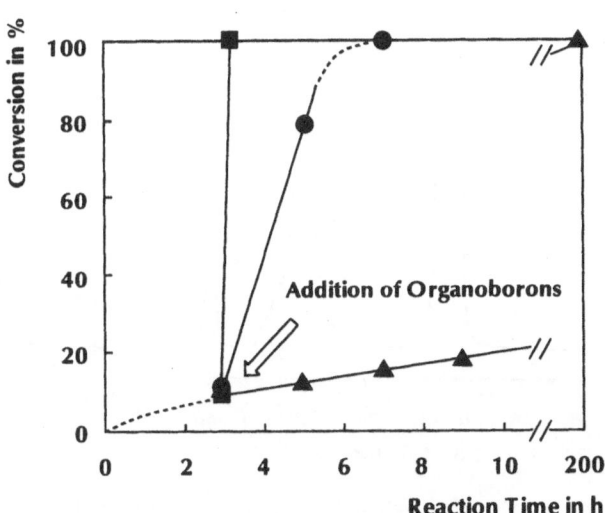

Fig. 13. Polymerization of methyl methacrylate (MMA) initiated with (TPP)AlMe (1, X= Me); $[MMA]_0/[1 \ (X=Me)]_0=200$, $[1 \ (X=Me)]_0=16.2$ mM, C_6H_6 as solvent, rt. Effects of triphenylboron (●), tris(pentafluorophenyl)boron (■), and tri-*n*-butylboron (▲) on the rate of polymerization; $[organoboron]_0/[2 \ (R=Me)]_0=1.0$

where the conversion of MMA (4.2% in the beginning) was increased to 19.7, 42.2, and 66.0%, respectively, in 20, 44, and 67 min. This corresponds to approximately a 40-times acceleration.

The GPC curve of the polymer formed at 100% conversion ($[(C_6H_5)_3B]_0/[2]_0=$ 1.0) was unimodal and very sharp (Fig. 14), where the Mw/Mn value was 1.04 and the degree of polymerization of the polymer (220) was very close to the initial monomer-to-initiator mole ratio of 200. Sequential two-stage polymerization of MMA using $(C_6H_5)_3B$ as the Lewis acid clearly demonstrated the living character of the polymerization. After the first-stage polymerization of MMA in the presence of $(C_6H_5)_3B$, a second feed of MMA was charged to the system. The GPC profile of this two-stage polymerization showed a clear increase in the Mn of the polymer, retaining a narrow molecular-weight distribution. These results clearly show a very long lifetime of the growing enolate species 2, even in the presence of $(C_6H_5)_3B$.

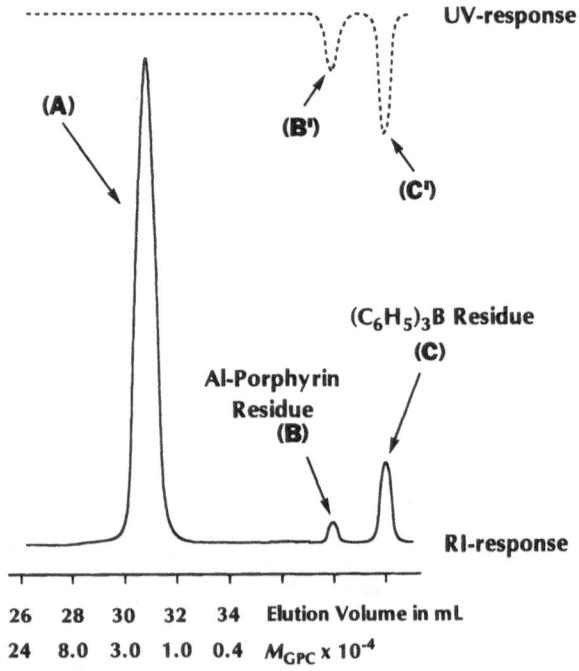

Fig. 14. Polymerization of methyl methacrylate (MMA) with the enolatealuminum porphyrin (2, R=Me)–triphenylboron $[(C_6H_5)_3B]$ system; (TPP)AlMe (1, X=Me) as initiator ($[MMA]_0/[1 (X=Me)]_0=200$, $[1 (X=Me)]_0=16.2$ mM), C_6H_6 as solvent, rt, $(C_6H_5)_3B$ added ($[(C_6H_5)_3B]_0/[2 (R=Me)]_0=1.0$) after irradiation for 3 h (9.3% conversion). GPC profiles (THF as eluent) of the polymerization mixture at 100% conversion (4 h), monitored with differential refractometer and UV (269 nm) detectors

Tris(pentafluorophenyl)boron [$(C_6F_5)_3B$], a triarylboron bearing electron-withdrawing perfluorinated phenyl rings, was found to be much more powerful than $(C_6H_5)_3B$ as a Lewis acid accelerator for the present polymerization. When $(C_6F_5)_3B$ was added at room temperature to the polymerization system at an equimolar ratio of $(C_6F_5)_3B$ to the growing species 2, the polymerization took place rapidly with considerable heat evolution, attaining 100% monomer conversion within only about 10 min [Fig. 13 (■)]. This polymerization was estimated to be 150-times faster than that in the absence of $(C_6F_5)_3B$, and 12.5-times faster than that with $(C_6H_5)_3B$ as a Lewis acid under similar conditions.

Unlike triarylborons, a trialkylboron, such as tributylboron (Bu$_3$B), neither accelerates nor terminates the polymerization under similar conditions [Fig. 13 (▲)]. On the other hand, the molecular-weight distribution of the produced polymer is very narrow, and the Mn value is close to the theoretical one. This is in sharp contrast to trialkylaluminums, which rapidly react with the growing species 2 to terminate the polymerization. Thus, no termination in the presence of Bu$_3$B indicates the essential difference in the susceptibility between the boron-carbon and aluminum–carbon bonds towards nucleophiles.

2.2.2
NMR Studies

In order to investigate the interaction of MMA with organoboron compounds, ^{13}C NMR studies were made. In the ^{13}C NMR spectrum in CD_2Cl_2 at 25 °C of an equimolar mixture of MMA and $(C_6F_5)_3B$ (Fig. 15), clear downfield shifts were observed for the signals of MMA due to $C=O$ [168.3 (a)→170.1 (a')], CH_2=[125.6 (c)→127.2 (c')], and CH_3O [52.2 (d)→53.1 (d')]. When the mole ratio of $(C_6F_5)_3B$ to MMA was increased from 1.0 to 5.0, the downfield shifts became more remarkable, particularly for the signal due to $C=O$ ($\Delta\delta$=3.4 ppm) [Fig. 16 (■)]. Thus, a coordination interaction exists between the carbonyl group of MMA and the boron atom of $(C_6F_5)_3B$. In the case where $(C_6H_5)_3B$ was present in place of $(C_6F_5)_3B$ under similar conditions, the downfield shifts of the MMA signals were again observed (●), but they were less explicit than those observed for the MMA–$(C_6F_5)_3B$ system (■). For example, the magnitude of the downfield shift for the MMA $C=O$ signal was only 0.1 ppm even at the mole ratio [$(C_6H_5)_3B$]/[MMA] of 5. On the other hand, in the MMA–Bu$_3$B system, the change in the chemical shifts of the MMA signals vs. the mole ratio of Bu$_3$B to MMA was much smaller than the above two triarylboron cases, where the MMA $C=O$ signal, for example, shifted only by 0.1 ppm when the mole ratio of Bu$_3$B to MMA was increased to 10 (▲). When these results are compared with those for two representative MMA–organoaluminum {MeAl[OC_6H_2(2,4,6-tri-tert-Bu)]$_2$, $(C_6H_5)_3Al$} systems, the extents of the downfield shifts for the MMA $C=O$ signal are increased in the order: Bu$_3$B (▲)<$(C_6H_5)_3Al$ (○)~$(C_6H_5)_3B$ (●)<$(C_6F_5)_3B$ (■)<MeAl[OC_6H_2(2,4,6-tri-tert-Bu)]$_2$ (□) (Fig. 16). This order roughly corresponds to the order of the observed acceleration effects of these Lewis acids for the 2-mediated polymerization [28].

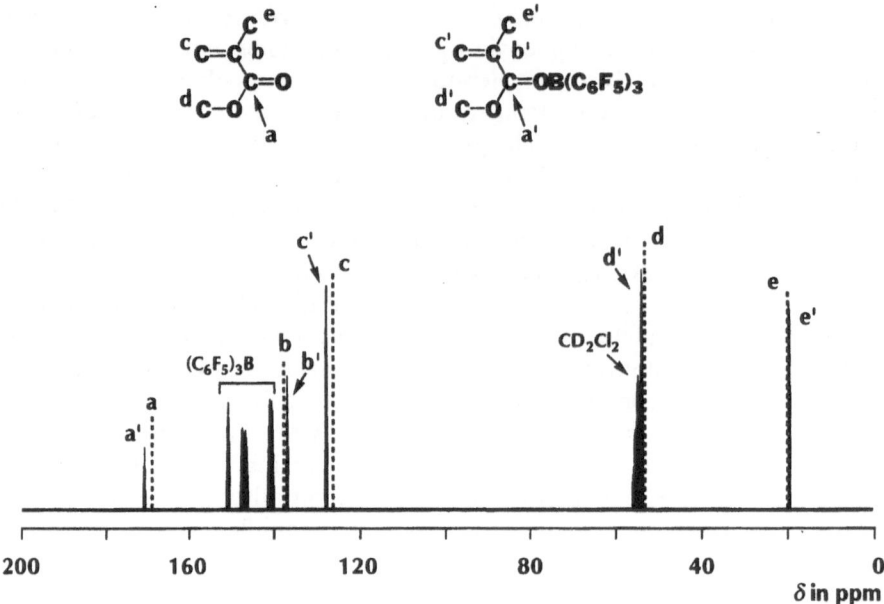

Fig. 15. ^{13}C NMR profile in CD$_2$Cl$_2$ at 25 °C of an equimolar mixture of methyl methacrylate (MMA) and tris(pentafluorophenyl)boron [(C$_6$F$_5$)$_3$B]. *Broken lines* represent the signals due to MMA in the absence of (C$_6$F$_5$)$_3$B

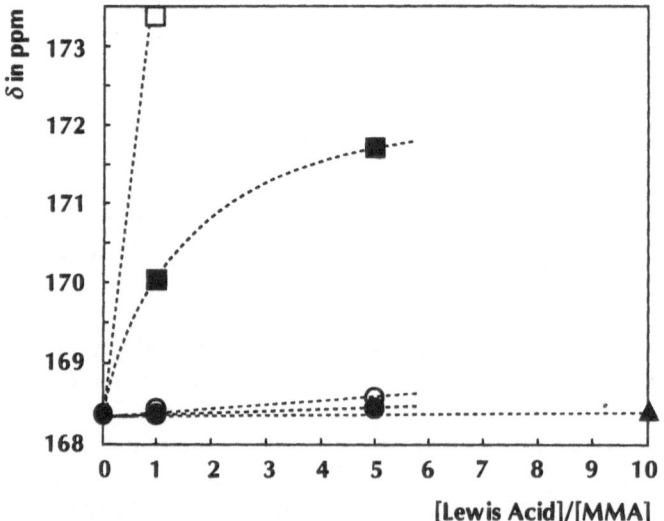

Fig. 16. ^{13}C NMR profiles in CD$_2$Cl$_2$ at 25 °C of methyl methacrylate (MMA)–Lewis acid mixtures. Relationships between the chemical shift of the *C*=O signal of MMA and the mole ratio of Lewis acid to MMA: Lewis acids; triphenylboron (●), tris(pentafluorophenyl)boron (■), tri-*n*-butylboron (▲), triphenylaluminum (○), and methylaluminum bis(2,4,6-tri-*tert*-butylphenolate) (□)

In relation to the living character of polymerization with the 2 (R=Me)–organoboron systems, a C_6D_6 solution of $(C_6H_5)_3B$ was added to an equimolar amount of the living polymer of MMA carrying an enolate reactive terminal (2, R=Me), prepared with $[MMA]_0/[1\ (X=Me)]_0$ of 10 at 100% conversion, and the mixture was studied by ^1H NMR at 30 °C. The living polymer 2 alone in C_6D_6 shows a characteristic signal at δ 0.53–0.62 ppm due to the MeO group in the terminal enolate unit attached to the aluminum porphyrin moiety [29a], while $(C_6H_5)_3B$ in C_6D_6 provides a set of signals at δ 7.88–7.36 ppm. If the transmetallation takes place between 2 (R=Me) and $(C_6H_5)_3B$, these signals should disappear, while new sets of signals due to 1 (X=Ph) [Al-Ph: δ 2.73 (o-H, d), 5.83 (m-H, m), and 6.08 ppm (p-H, t)] [30] and the corresponding boron enolate species should appear. However, throughout the observation over a period of 24 h at 25 °C, the signals due to the starting aluminum enolate species 2 (R=Me) and $(C_6H_5)_3B$ remained unchanged in their relative intensities to the pyrrole-β-protons of the aluminum porphyrin moiety (δ 9.37 ppm, 8H, s), and no new signals characteristic of 1 (X=Ph) appeared [30]. This observation is in agreement with the result of the two-stage polymerization of MMA with the 2 (R=Me)–$(C_6H_5)_3B$ system, where the livingness of polymerization succeeded perfectly from the first to the second stage even when the long interval (48 h) was placed before the second monomer feed. More surprisingly, no ligand-exchange reaction took place when the mixture of 2 and $(C_6H_5)_3B$ was heated at 60 °C for 3 h.

2.3
"One-Shot" Lewis Acid Promoted Living Polymerization of Methyl Methacrylate [31]

For Lewis acid promoted living polymerization of MMA with (TPP)AlMe (1, X=Me) as initiator, a photoinitiation prior to the addition of the Lewis acid is required. This is because (1) 1 (X=Me) without irradiation does not have the ability to initiate the polymerization even in the presence of Lewis acid, and (2) "all-at-once" polymerization by direct irradiation of a mixture of MMA, 1 (X=Me), and the Lewis acid results in the formation of a relatively broad MWD PMMA with Mn much higher than expected. In this sense, the procedure using 1 (X=Me) as initiator is not convenient for practical application. In this section, we report on aluminum porphyrins with various axial ligands which were tested as initiators in order to realize a more convenient, "one-shot" high-speed living polymerization of methyl methacrylate with no need for irradiation with visible light.

Aluminum porphyrins 1 having axial ligands such as chloride (X=Cl), acetate (X=O_2CMe), phenolate [X=OC_6H_3(2,4-di-$tert$-Bu)], methanolate (X=OMe), propanethiolate (X=SPr), and benzenethiolate (X=SPh) were tested as initiators for the polymerizations of MMA (200 equiv with respect to 1) in the absence and presence of methylaluminum bis(2,6-di-$tert$-butyl-4-methylphenolate) (3e) as Lewis acid under diffuse light in CH_2Cl_2 at room temperature. As a result, the aluminum porphyrins carrying aluminum–chlorine (1, X=Cl) and –oxygen bonds [1, X=O_2CMe, OC_6H_3(2,4-di-$tert$-Bu), OMe] did not bring about the po-

lymerization at all in the absence and even in the presence of the Lewis acid **3e**.
On the contrary, aluminum porphyrins 1 bearing aluminum–sulfur bonds (X=
SPr, SPh) were found to initiate the polymerization without irradiation [32] even
in the absence of **3e**.

Thus, one-shot high-speed living polymerization of MMA without a photoinitiation step is possible by the direct addition of a mixture of MMA and a Lewis
acid **3e** to the initiator solution [33]. For example, when 100 equiv of MMA containing 1.5 mol% of **3e** were added to a C_6H_6 solution of (TPP)AlSPr (**1**, X=SPr),
the color of the solution immediately turned from reddish brown to reddish purple characteristic of an enolatealuminum porphyrin species. The polymerization took place rapidly with heat evolution, and was complete within only 90 s,
affording the polymer with Mn and Mw/Mn, as estimated by GPC, of 11,200 and
1.13, respectively (Fig 17A). An excellent agreement between the thus estimated
Mn value and that expected from the monomer-to-initiator mole ratio (10,000)
indicated the participation of all the molecules of (TPP)AlSPr in initiating the
high-speed polymerization. When MMA was added again after the first-stage
polymerization was complete, second-stage polymerization ensued to allow the

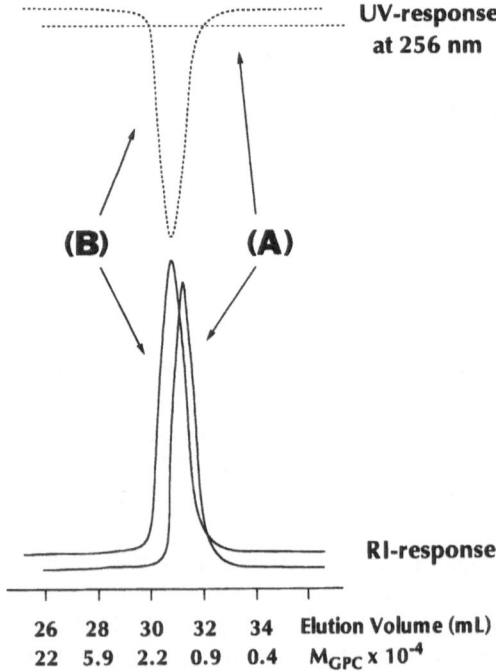

Fig. 17. GPC profiles of poly(methyl methacrylate)s obtained with *A* The (TPP)AlSPr (**1**, X=
SPr)–methylaluminum bis(2,6-di-*tert*-butyl-4-methylphenolate) (**3e**) and *B* The (TPP)Al-
SPh (**1**, X=SPh)–**3e** system. $[MMA]_0/[1]_0/[3e]_0=100/1.0/1.0$, C_6H_6 as solvent, rt, 100% con-
version

quantitative formation of a poly(MMA) with narrow molecular-weight distribution (living polymerization).

From an end-group analysis of the polymer by ^{1}H and ^{13}C NMR spectroscopy and GPC monitoring RI and UV responses, the polymerization of MMA with the (porphyrinato)aluminum thiolate–Lewis acid (**3e**) system was initiated by a nucleophilic attack of the thiolate group of the initiator to the monomer, leading to the formation of an alkyl- or arylthio moiety at the polymer terminal. The ^{13}C NMR spectrum (10–40 ppm) in CDCl$_3$ of the polymer synthesized by the **1** (X= SPr)-**3e** system (Mn= 6,900; Mw/Mn=1.13) showed a set of weak signals assignable to the propylthio group attached to the polymer terminal [a, δ 13.4 (CH$_3$); b, δ 23.2 (CH$_2$CH$_3$); c, δ 35.3 (SCH$_2$CH$_2$CH$_3$)] [34] and that due to the methylene group of the terminal MMA unit attached to the -SPr group [d, δ 36.3 ppm (CH$_2$SPr)] was observed in addition to the signals due to the methyl group in the polymer main chain (e, δ 18.8 and 16.4) (Fig. 18) [35]. The ^{1}H NMR spectrum also showed the signals attributable to the terminal structure -C(CH$_3$) (CO$_2$CH$_3$)CH$_2$SPr (SCH$_2$Et, δ 2.41; CH$_2$SPr, δ 2.51 and 2.75) [35], and the number-average degree of polymerization, as calculated from the intensity ratio of the signal due to the main chain CH$_2$ group (δ 1.4–2.1) [35] to that due to SCH$_2$Et, was 73, which is close to the value estimated by GPC (69). Thus, every terminal of the polymer molecule carries a propylthio group originating from the initiator **1** (X=SPr). Incorporation of the thiolate group of the initiator into the polymer terminal was also demonstrated for the polymerization with the (TPP)AlSPh (**1**, X=Ph)–**3e** system. As indicated by the GPC chromatogram, the

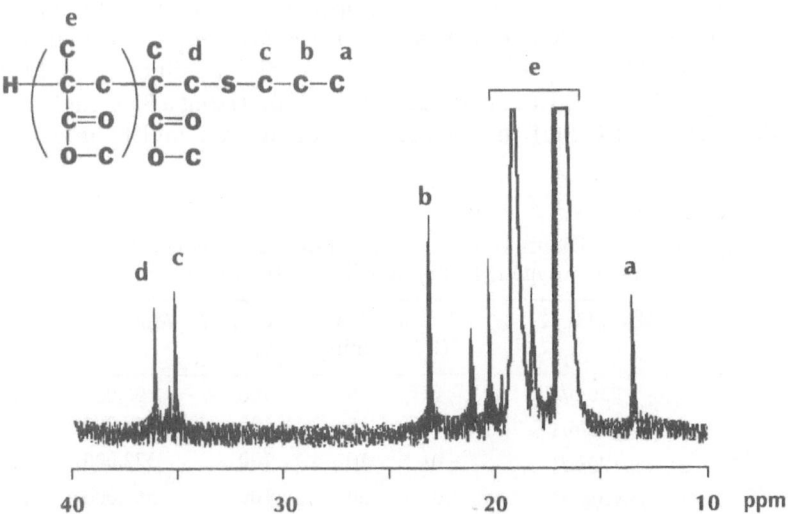

Fig. 18. ^{13}C NMR spectrum in CDCl$_3$ of the poly(methyl methacrylate) (Mn_{GPC}=6900, Mw/Mn=1.13) obtained with the (TPP)AlSPr (**1**, X=SPr)–methylaluminum bis(2,6-di-*tert*-butyl-4-methylphenolate) (**3e**) system. [MMA]$_0$/[**1**]$_0$/[**3e**]$_0$=30/1.0/0.5, 100% conversion

polymer, purified by repeated precipitations, exhibits a clear UV response when monitored at 256 nm [38]. On the other hand, the polymers prepared with the systems 1 (X=Me)–3e and 1 (X=SPr)–3e were both silent at 256 nm [37]. Thus, the clear UV response observed for the polymer formed with the 1 (X=SPh)–3e system [(B) in Fig. 17] indicates the presence of an aromatic sulfide moiety at the polymer terminal.

2.4
Synthesis of Monodisperse, High-Molecular-Weight Poly(methyl Methacrylate) [39]

The synthesis of monodisperse, ultrahigh-molecular-weight polymers is one of the most challenging subjects of both fundamental and practical interest. The aluminum porphyrin 1–Lewis acid systems can be used for this purpose by taking advantage of their strikingly high activities (Table 6) [40]. A typical example is given below. To a CH_2Cl_2 solution (12.0 ml) of (TPP)AlMe (1, X=Me; 0.30 mmol) in a 200-ml round-bottomed flask carrying a side ampule (20 ml) (Fig. 19) was added MMA (30 mmol) with a syringe; the mixture was irradiated with xenon arc light (λ>420 nm) at 35 °C for 4 h, whereupon 1 (X=Me) was completely converted to the aluminum enolate species2 (R=Me). Then, the irradiation was stopped, and the flask was tilted to the ampule side in order to minimize the amount of 2 (R=Me) remaining in the flask to provide a high MMA-to-2 mole ratio in the next stage polymerization. After most of the initiator solution had been transferred to the side ampule, a CH_2Cl_2 solution of a mixture of 3590 molar amount of MMA and 85 molar amount of 3e (2.5 mol% with respect to MMA) was added to the remaining solution with a syringe; the polymerization was conducted at 0 °C. Complete consumption of MMA was attained within 20 min, affording a polymer with Mn and Mw/Mn, respectively, of 5.6×10^5= 560,000 and 1.1. Similarly, PMMA with Mn exceeding a million ($Mn=1.02\times10^6$= 1,017,000, Mw/Mn=1.2) was synthesized by polymerizing a 8300 molar amount of MMA with 2 (R=Me) in the presence of 3e (4.2 mol%) under similar conditions.

Table 6. Polymerization of methyl methacrylate (MMA) via an enolatealuminum porphyrin (2, R=Me) in the presence of methylaluminum diolates (MeAlX$_2$)[a]

Run	3	$[MMA]_0/[3]_0[2]_0$	Temp/ °C	Time/ min	Conv.[b]/ %	Mn[c]	Mw/Mn[c]
1	3b	930/3/1	−40	5	100	96,000	1.1
2	3b	2,160/6/1	−40	15	100	176,000	1.1
3	3b	3,170/8/1	−40	10	100	372,000	1.2
4	3e	3,590/85/1	0	20	100	560,000	1.1
5	3e	8,300/350/1	0	35	100	1,017,000	1.2

[a]   Polymerization was carried out in CH_2Cl_2 under nitrogen.
[b]   Determined by ^{1}H NMR analysis of the reaction mixture.
[c]   Estimated by GPC based on polystyrene standards [14].

2.5
Accelerated Living Polymerization of Methacrylonitrile with Aluminum Porphyrin Initiators by the Activation of Monomer or Growing Species [39]

For the anionic polymerization of methacrylonitrile (MAN), many initiators have been developed, which include alkali-metal alkyls such as butyllithium [42], triphenylmethylsodium [43], phenylisopropylpotassium [43], the disodium salt of living α-methylstyrene tetramer [44], alkali-metal amides [45], alkoxides [46], and hydroxide [47], alkali metal in liquid NH_3 [48], quaternary ammonium hydroxide [49], and a silyl ketene acetal coupled with nucleophilic or Lewis acidic catalysts [50]. However, only a single example of the synthesis of PMAN with narrow molecular-weight distribution can be cited, and the reported number-average molecular weights were much higher than those calculated from the stoichiometry of the butyllithium initiator [42].

As mentioned above, the new method "Lewis acid promoted living polymerization" of methacrylic esters, by using enolatealuminum porphyrin (2) as nucleophilic initiator in conjunction with organoaluminum compounds, such as methylaluminum bis(2,6-di-*tert*-butyl-4-methylphenolate) (3e), as Lewis acids has enabled us to synthesize poly(methacrylic ester) of narrow molecular-weight distribution [51]. On the other hand, some reactions of aluminum por-

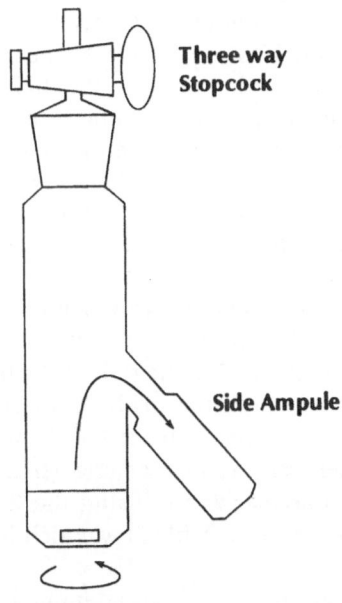

Three way Stopcock

Side Ampule

Magnetic Stirring

Fig. 19. The polymerization flask designed for the synthesis of high-molecular-weight ($Mn>500,000$) poly(methyl methacrylates) by the high-speed living polymerization with the (porphyrinato)aluminium enolate (2)–Lewis acid (3e) systems

phyrin take place only when Lewis bases are added to the reaction system. For example, the addition of a Lewis base is essential for the reaction of aluminum porphyrins with carbon dioxide. It is considered that the added Lewis base coordinates to the central aluminum atom and enhances the reactivity of aluminum porphyrins [52].

In the present section we describe the living anionic polymerization of methacrylonitrile by two initiating systems such as the aluminum porphyrin–Lewis acid system and the aluminum porphyrin–Lewis base system which enables the synthesis of poly(methyl methacrylate-*b*-methacrylonitrile)s of controlled molecular weights.

2.5.1
Polymerization of Methacrylonitrile with Methylaluminum Porphyrin in the Presence of Methylaluminum Diphenolate

Polymerization of methacrylonitrile (MAN) was first attempted by using methylaluminum porphyrin (**1**, X=Me) as initiator under diffuse light in CH_2Cl_2 at room temperature ($[MAN]_0/[1]_0$=100); however, the reaction of MAN and **1** did not occur at all. On the other hand, under irradiation with visible light (λ>420 nm) at 35 °C, the polymerization of MAN took place, accompanied by a color change from bluish purple to brownish purple. The polymerization proceeded very slowly, and the monomer conversion reached only 25% even after one week, affording polymethacrylonitrile (PMAN) with Mn and Mw/Mn values of 2900 and 1.31, respectively. In sharp contrast, when **3e** (10 equiv with respect to **1**) was added to the above reaction mixture, the chain growth readily proceeded and was complete in 5 h. The polymerization system was homogeneous at the earlier stage of the reaction, but turned heterogeneous at the later stage due to the poor solubility of PMAN in the polymerization medium. The GPC profile of the finally obtained PMAN showed a unimodal but broad chromatogram (Mw/Mn=1.35). The Mn of the polymer (7800), as estimated from GPC, was close to that expected from the initial monomer-to-initiator mole ratio by assuming that every molecule of aluminum porphyrin produced one polymer molecule (6700), indicating that the added **3** could not be a new initiator for the polymerization of MAN. Actually, **3** alone did not bring about the polymerization of MAN under identical conditions. ^1H NMR in $CDCl_3$ at 55 °C of poly(methyl methacrylate) (PMMA) converted from PMAN produced as above [53] indicated that the polymer was almost atactic (I/H/S=0.28/0.52/0.20), while PMMA prepared by the polymerization using the **2–3e** system (Mn=10,700, Mw/Mn=1.19; prepolymer in no. 5, Table 1) was rich in syndiotactic sequences (I/H/S=0.02/0.37/0.61).

The block copolymerization of MAN was attempted from a living prepolymer of MMA carrying a (porphinato)aluminum enolate growing terminal (Scheme 6); however, without **3e**, the polymerization of MAN from **2** did not occur. However, when 100 equiv of MAN were added to a CH_2Cl_2 solution of **2** {Mn=12,000, Mw/Mn=1.11; prepared with $[MMA]_0/[1]_0$ of 100, 100% conver-

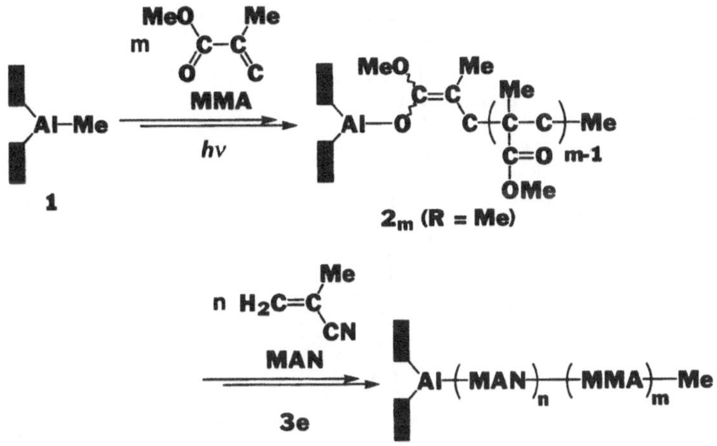

Scheme 6

sion, under irradiation ($\lambda > 420$ nm); Fig. 20 (I)} in the presence of **3e** ($[3e]_0/[2]_0 = 10.0$) at room temperature, the polymerization of MAN started with a color change of the polymerization system. The polymerization proceeded smoothly and was complete within 3 h, where the polymerization mixture was homogeneous throughout the reaction.

The GPC profile provided a unimodal, sharp elution pattern for the finally obtained polymer, in which the peak due to the prepolymer of MMA was not observed, showing that a MMA–MAN block copolymer was produced quantitatively [Fig. 20 (II)]. The Mn of the polymer, as estimated by GPC, was 20,800 ($Mw/Mn = 1.17$), which is close to that (16,700) expected from the initial MAN-to-**2** mole ratio. Successful block copolymerization indicates that all the molecules of the growing PMMA (**2**) produced in the first stage participated in initiating the second-stage polymerization of MAN.

When the polymerization of MAN using the living PMMA as initiator was carried out in the presence of **3e** (3 equiv with respect to **2**) at room temperature, the Mn of the polymer increased linearly with the conversion, which is in good agreement with the value expected from the initial mole ratio of MAN to **2** and the conversion (Fig. 21), and the molecular-weight distribution (MWD) of the block copolymer stayed narrow ($Mw/Mn = 1.1$–1.2) throughout the polymerization.

By increasing the amount of **3e** from 3.0 to 10.0 equiv with respect to **2**, the polymerization of MAN proceeded more rapidly (Fig. 22). However, the Mn of the PMAN sequence obtained at 100% conversion was almost constant irrespective of the ratio of **3e** to **2** (Table 7, runs 2–4) These results indicate that the added **3** did not serve as initiator but served as accelerator for the polymerization of MAN.

When the initial mole ratio of MAN to the prepolymer of MMA (**2**) was increased from 50 to 200, the Mn of the PMAN sequence was increased almost proportionally from 5400 to 16,400 retaining the ratio Mw/Mn in the range of 1.2–1.3 (Table 7, runs 5–7, 10; Fig. 23). Since PMMA–PMAN block copolymers with

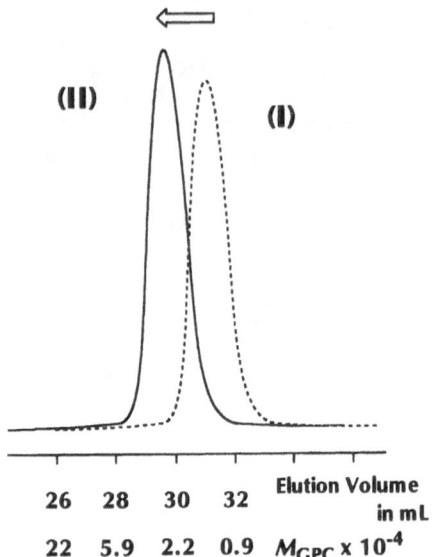

Fig. 20. Block copolymerization of methyl methacrylate (MMA) and methacrylonitrile (MAN) in CH_2Cl_2 with the methylaluminum porphyrin (**1**, X=Me)–methylaluminum bis(2,6-di-*tert*-butyl-4-methylphenolate) (**3e**) system. GPC profiles of *I* The prepolymer of MMA ([MMA]$_0$/[1]$_0$=100, under irradiation λ>420 nm, 35 °C, 100% conversion; *Mn*= 12,000, *Mw*/*Mn*=1.13) and *II* The block copolymer of MMA and MAN (in the presence of **3e**, [MAN]$_0$/[1]$_0$/[3e]$_0$=100/1.0/10, rt, 100% conversion; *Mn*=20,800, *Mw*/*Mn*=1.17)

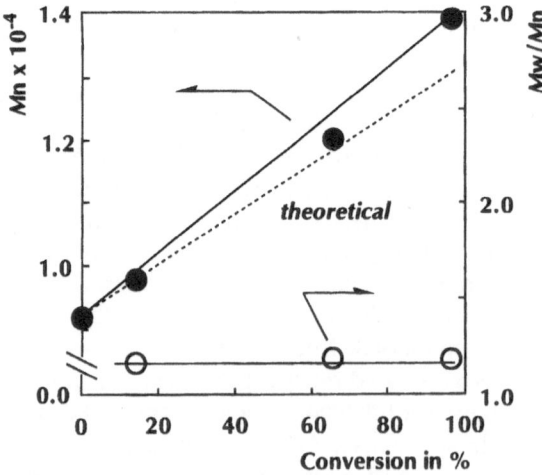

Fig. 21. Polymerization of methacrylonitrile (MAN) with the living prepolymer of methyl methacrylate (MMA) (**2**, m=50; *Mn*=9200, *Mw*/*Mn*=1.13)–methylaluminum bis(2,6-di-*tert*-butyl-4-methylphenolate) (**3e**) system; [MAN]$_0$/[2]$_0$/[3e]$_0$=50/1.0/3.0, [2]$_0$=22.6 mM, CH_2Cl_2 as solvent, rt. Relationship between *Mn* (●) [*Mw*/*Mn* (○)] of the polymer and conversion

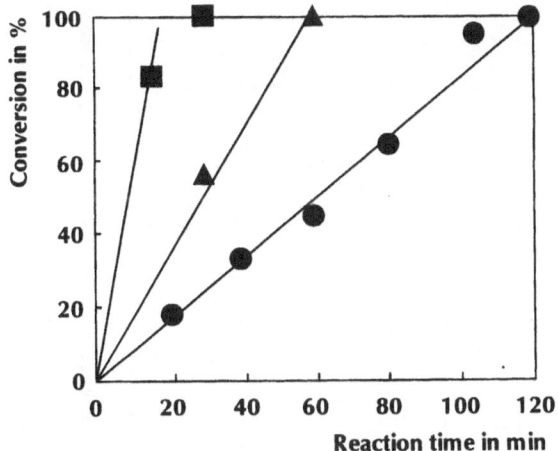

Fig. 22. Polymerization of methacrylonitrile (MAN) with the living prepolymer of methyl methacrylate (MMA) (2)–methylaluminum bis(2,6-di-*tert*-butyl-4-methylphenolate) (3e) system; $[MAN]_0/[2]_0=50$, $[2]_0=22.6$ mM, CH_2Cl_2 as solvent, rt, initial ratios of 3e to 2=3.0 (●), 4.0 (▲), and 10 (■). Effect of the amount of Lewis acid 3e on the rate of polymerization

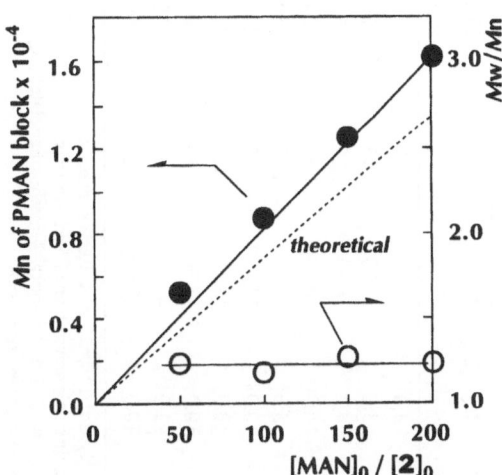

Fig. 23. Polymerization of methacrylonitrile (MAN) with the living prepolymer of methyl methacrylate (MMA) (2)–methylaluminum bis(2,6-di-*tert*-butyl-4-methylphenolate) (3e) system in CH_2Cl_2 at rt. Relationship between Mn of the PMAN block (●) [Mw/Mn (○)] of the polymer formed at 100% conversion and the initial monomer-to-initiator mole ratio ($[MAN]_0/[2]_0$)

Table 7. Block copolymerization of methacrylonitrile (MAN) with the living prepolymer of methyl methacrylate (MMA) (2) in the presence of methylaluminum bis(2,6-di-tert-butyl-4-methylphenolate) (3e)[a]

Run	$[MMA]_0/[1]_0$	$[3e]_0/[1]_0$	Prepoymer (2)[b] Mn^c	Mw/Mn^c	$[MAN]_0/[2]_0$	Time/h	Block copolymer Mn^c	Mw/Mn^c	Mn of PMAN	Block[d]	mol% of MAN[e]
1	30	4.0	–	–	100	18	11,100	1.27	–	(6,700)[f]	77.3
2	50	3.0	9,200	1.13	50	2	16,000	1.19	6,800	(3,400)	
3	50	4.0	6,500	1.27	50	1	11,700	1.24	5,200	(3,400)	
4	50	10.0	6,600	1.37	50	0.5	11,200	1.35	4,600	(3,400)	
5	100	3.0	10,700	1.19	50	2	16,100	1.23	5,400	(3,400)	40.1
6[g]	100	10.0	11,200	1.11	100	3	20,800	1.17	9,600	(6,700)	56.7
7[g]	100	10.0	11,100	1.19	150	3	23,600	1.35	12,500	(10,100)	62.3
8[g]	100	10.0	13,500	1.14	200	5	24,000	1.48	10,500	(13,400)	
9	200	3.0	–	–	50	3	30,300	1.18	–	(3,400)	19.4
10[h]	200	10.0	22,900	1.11	200	14	39,300	1.24	16,400	(13,400)	50.1

[a] In CH_2Cl_2 under N_2 at rt, $[2]_0$=25 mM, 100% conversion.
[b] Prepared by the living polymerization of MMA initiated with (TPP)AlMe (1) in CH_2Cl_2 at 35 °C under irradiation with xenon arc light (λ>420 nm) in the presence of methylaluminum bis(2,6-di-tert-butyl-4-methylphenolate) (3e), 100% conversion.
[c] By GPC based on polystyrene standards.
[d] (Mn of the block copolymer)/(Mn of the prepolymer).
[e] Calculated from the N content in elemental analysis.
[f] Calculated from $\{[MAN]_0/[2]_0 \times MW$ of MAN [67]}.
[g] $[2]_0$=12.5 mM.
[h] $[2]_0$=6.3 mM.

longer PMMA sequences are more soluble, the polymerization mixture was homogeneous throughout the polymerization of MAN even when a block copolymer with a long PMAN sequence was formed, and the MWD of the produced block copolymer was narrow.

The polymerization of MAN with **2** in the presence of **3e** proceeded with living character. Clear evidence for this was given by the sequential two-stage polymerization of MAN initiated with enolatealuminum porphyrin **2** [living PMMA; $Mw=13,200$; $Mw/Mn=1.22$; Fig. 24 (I)] in the presence of **3** (1:3), where 50 equiv of MAN with respect to **2** were polymerized up to 100% conversion (4 h) at the first stage, and after a 16-h interval at room temperature, 50 equiv of MAN were charged to the mixture. In spite of the long interval between the completion of the first-stage polymerization and the second monomer feed, the second-stage polymerization ensued, and was complete within 6 h. The GPC profile of the product (the block copolymer) showed a clear increase in molecular weight from 17,500 to 20,700 with the shift of the sharp peak (Mw/Mn: from 1.21 to 1.29) [Fig. 24 (II) and (III)].

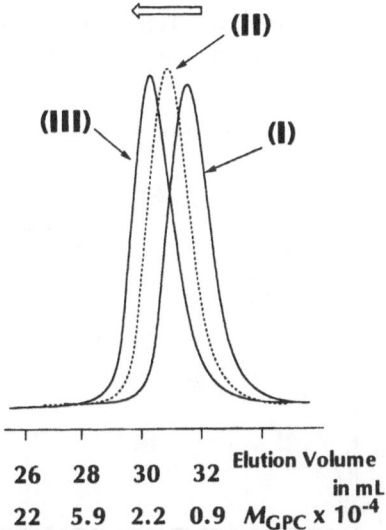

Fig. 24. Sequential three-stage polymerization of methyl methacrylate (MMA), methacrylonitrile (MAN), and MAN in CH_2Cl_2 at rt with the methylaluminum porphyrin (1, X=Me)–methylaluminum bis(2,6-di-*tert*-butyl-4-methylphenolate) (3e) system. GPC profiles of the polymers formed at *I* The first stage $[MMA]_0/[1]_0=100$, 100% conversion; $Mn=13,200$, $Mw/Mn=1.22$, *II* The second stage $[MAN]_0/[1]_0=50$, 100% conversion; $Mn=17,500$, $Mw/Mn=1.21$, and *III* The third stage $[MAN]_0/[1]_0=50$, 100% conversion; $Mn=20,700$, $Mw/Mn=1.29$

2.5.2
Polymerization of Methacrylonitrile with Methylaluminum Porphyrin in the Presence of Pyridine

Pyridine in place of methylaluminum bis(2,6-di-*tert*-butyl-4-methylphenolate) (3e) was also effective for the polymerization of MAN from the living PMMA (2). An example is shown by the polymerization of MAN under irradiation with the 2 (Mn=4000, Mw/Mn=1.09; prepared with $[MMA]_0/[1]_0$=40, 100% conversion)–pyridine system at the initial mole ratio $[MAN]_0/[2]_0/[pyridine]_0$ of 100/1.0/0.6. MAN was polymerized up to 63% conversion in 65 h [Fig. 25 (□)], where the GPC peak of the polymer formed was observed to shift clearly towards the higher molecular-weight region (Mn=7600), retaining the narrow MWD (Mw/Mn=1.26), and the peak corresponding to the prepolymer of MMA was not observed. These facts clearly demonstrate the successful polymerization of MAN from 2, affording a PMMA–PMAN block copolymer. In sharp contrast to the polymerization of MAN, polymerization of MMA with aluminum porphyrin 2 was retarded by pyridine. For example, in the presence of 2 equiv of pyridine with respect to 2, the polymerization of 100 equiv of MMA proceeded very slowly to attain 25% conversion in 18 h under irradiation, while in the absence of pyridine, the polymerization of MMA with 2 was complete within 12 h under otherwise identical conditions.

By increasing the amount of pyridine from 0.6 to 10 equiv with respect to the living PMMA (2) ($[MAN]_0/[2]_0$=50), the polymerization of MAN under visible light irradiation proceeded more rapidly [Fig. 25 (○)], where the Mn of the produced polymer increased linearly with conversion, retaining the ratio Mw/Mn in

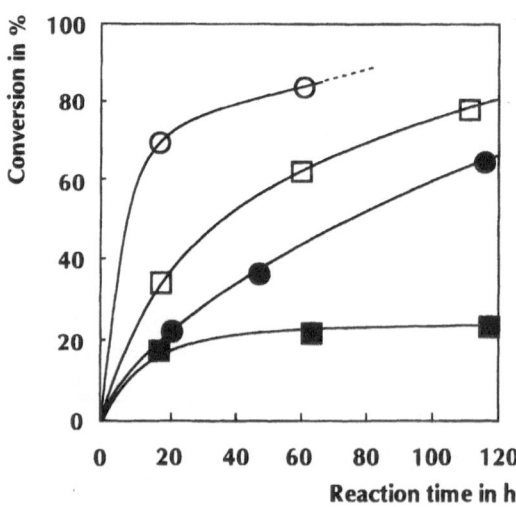

Fig. 25. Polymerization of methacrylonitrile (MAN) with the living prepolymer of methyl methacrylate (MMA) (2)–pyridine system; $[2]_0$=20.7 mM, CH_2Cl_2 as solvent, rt, $[MAN]_0/[2]_0$=100, $[pyridine]_0/[2]_0$=0.6 (■, □), 10 (●, ○), under diffuse light (■, ●), under irradiation of visible light (λ>420 nm) (□, ○). Time-conversion relationship

the range of 1.1–1.3. A good agreement between the observed and expected Mns of the polymer (solid and broken lines, respectively) again indicates the quantitative initiation from every molecule of the living PMMA (2).

It is particularly noteworthy that a dramatic acceleration by irradiation with visible light was observed in the presence of pyridine for the polymerization of MAN initiated with the living PMMA (2) (Fig. 25 ■→□, ●→○) [54]. For example, the polymerization of MAN with 2 ($[MAN]_0/[2]_0/[pyridine]_0=100/1.0/10$) proceeded rapidly up to 67% conversion in 18 h under visible light irradiation ($\lambda>420$ nm), while in diffused light under otherwise identical conditions, the conversion of MAN after 20 h was only 22%. All the block copolymers produced under irradiation and under diffused light were of narrow MWD.

2.6
Living Ring-Opening Polymerization of Heterocyclic Monomers with Aluminum Porphyrin–Organoaluminum Compound Systems

Aluminum porphyrins are excellent initiators for the living polymerizations of a wide variety of monomers such as epoxides [55], lactones [56], alkyl methacrylates [57], and alkyl acrylates [58] affording the corresponding polymers of controlled molecular weights with narrow molecular-weight distribution. In the course of the study mentioned above, it was found that the polymerization of methyl methacrylate with aluminum porphyrin, proceeding via an enolatealuminum porphyrin as the growing species [57], is dramatically accelerated by the addition of a bulky Lewis acid 3f, where the polymerization is complete within seconds under appropriate conditions to give a polymer with a narrow MWD [59]. The basic concept of this high-speed living polymerization involves the coordinative activation of the monomer by the bulky Lewis acid which does not react directly with the growing species on bulky aluminum porphyrin (Fig. 11). In the present section, the successful extension of the above method to the Lewis acid assisted anionic living ring-opening polymerizations with aluminum porphyrin is described for the polymerizations of epoxide, lactone, and oxetane. Furthermore, in order to generalize the basic concept of the Lewis acid assisted living polymerization, aluminum complexes of phthalocyanine (11), tetraazaannulene (12), and Schiff bases (13–15) were also employed as initiators.

Structure 11

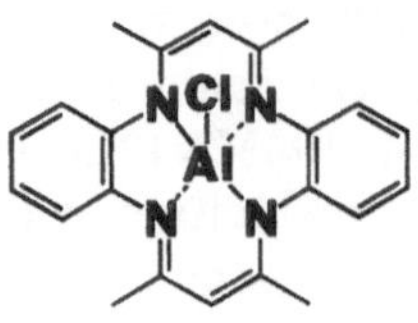

12

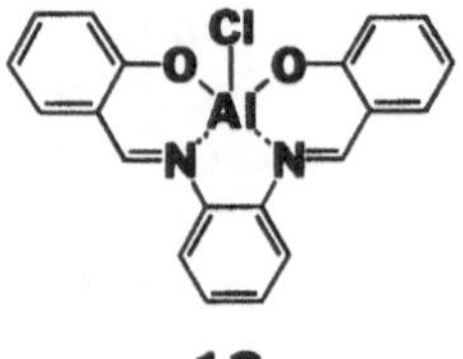

13

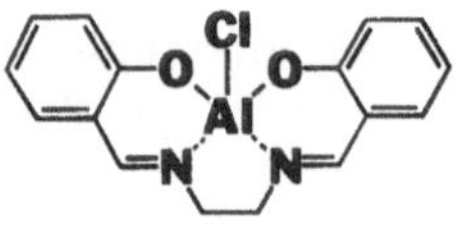

14

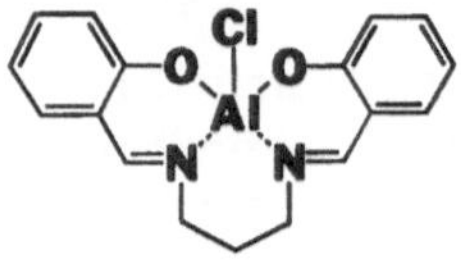

15

Structure 12-15

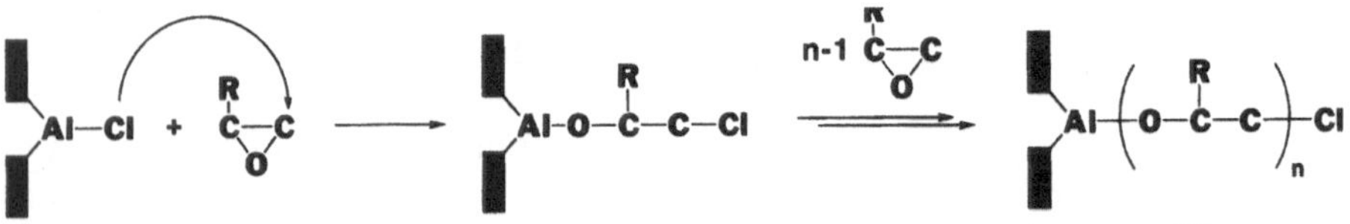

1 (X = Cl)   **16₁**   **16ₙ**

Scheme 7

2.6.1
Ring-Opening Polymerization of Epoxide Initiated with Aluminum Porphyrin in the Presence of an Organoaluminum Compound [60]

Polymerization of 1,2-epoxypropane (propylene oxide, PO), carried out at room temperature using chloroaluminum porphyrin [(TPP)AlCl; **1**, X=Cl] as initiator in CH_2Cl_2 at the initial mole ratio $[PO]_0/[1]_0$ of 200, proceeded, via alcoholatealuminum porphyrin as the growing species **16** (Scheme 7), rather slowly, to attain 19.8% conversion in 7 h, giving a polymer with Mn and Mw/Mn of 3300 and 1.05, respectively. On the other hand, when a CH_2Cl_2 solution of methylaluminum bis(2,4,6-tri-*tert*-butylphenolate) (**3f**) (0.25 mol% with respect to PO) was added to the reaction mixture, a rapid polymerization took place with vigorous heat evolution to attain 85.5% conversion within only 3 min. The extent of acceleration was estimated to be 460-times. The GPC profile of the polymer thus formed showed a unimodal, sharp chromatogram, from which the Mw/Mn ratio was estimated, based on polystyrene standards, to be 1.21. The Mn of the produced polymer was estimated to be 11,900, which is close to the expected value (9900), taking the monomer-to-initiator mole ratio and conversion into consideration.

Polymerization of PO using an alcoholatealuminum porphyrin [(TPP) Al(PO)$_{10}$Cl, **16**; n=10] as initiator also proceeded rapidly in the presence of **3f**: As soon as 200 equiv of PO were added to a CH$_2$Cl$_2$ solution of (TPP)Al(PO)$_{10}$Cl (**16**, n=10) containing only 0.1 mol% of methylaluminum diphenolate **3f** with respect to PO, the polymerization started and attained 52.5, 73.6, and 81.1% conversion in 1, 2, and 3 min, respectively (Fig. 26). The Mn value of the polymer increased linearly with conversion, retaining the ratio Mw/Mn at 1.1 (Fig. 27). An excellent agreement between the observed and expected Mns of the polymer again indicates the quantitative initiation and propagation from every molecule of the starting alcoholatealuminum porphyrin **16**.

The rate of polymerization by the **16** (n=10)–**3f** system is dependent on the concentration of **3f**: When the initial mole ratio of **3f** to PO was increased from 0.025 to 2.5 mol%, the polymerization was much more accelerated to attain 94.1% conversion in only 3 min (Fig. 28). If the added **3f** initiates the polymerization, the number of the polymer molecules produced should increase incrementally in relation to the amount of **3f**, and the Mn value should decrease. However, irrespective of the mole ratio of **3f** to PO (0.025–2.5 mol%), the observed Mn values at 100% conversion were all close to the theoretical Mn (11,600) as indicated by the ratio of the numbers of the molecules of polymer and **16** (N_p/N_{TPP}) (Fig. 29) [61] being almost constant at unity, and the produced polymers were all of narrow MWD. The Mn of the polymer could be controlled by

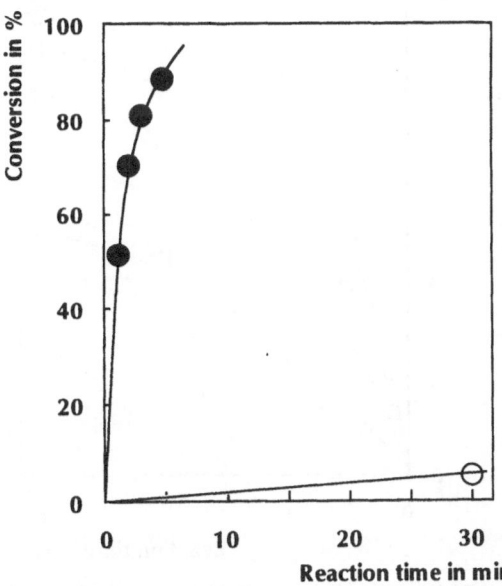

Fig. 26. Polymerization of 1,2-epoxypropane (PO) by the alcoholatealuminum porphyrin (**16**, n=10)–methylaluminum bis(2,4,6-tri-*tert*-butylphenolate) (**3f**) system, [PO]$_0$/[**16**]$_0$/ [**3f**]$_0$=200/0.2/1.0, [**3f**]$_0$=18.5 mM, CH$_2$Cl$_2$ as solvent, rt. Time-conversion curve

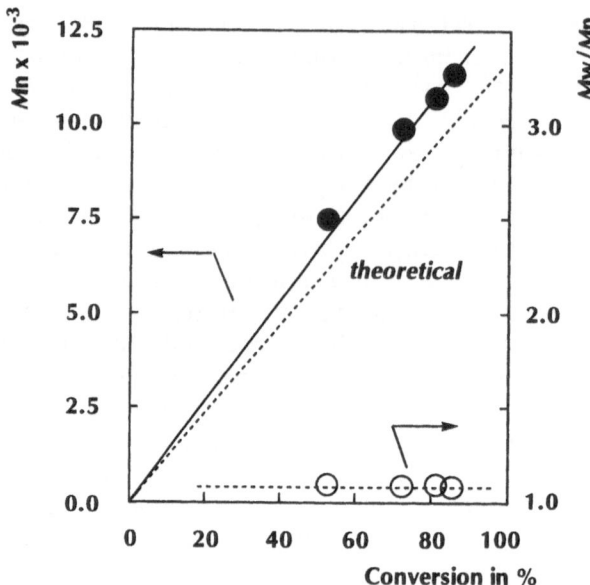

Fig. 27. Polymerization of 1,2-epoxypropane (PO) by the alcoholatealuminum porphyrin (**16**, n=10)–methylaluminum bis(2,4,6-tri-*tert*-butylphenolate) (**3f**) system. Relationship between Mn (●) [Mw/Mn (○)] of the polymer and conversion

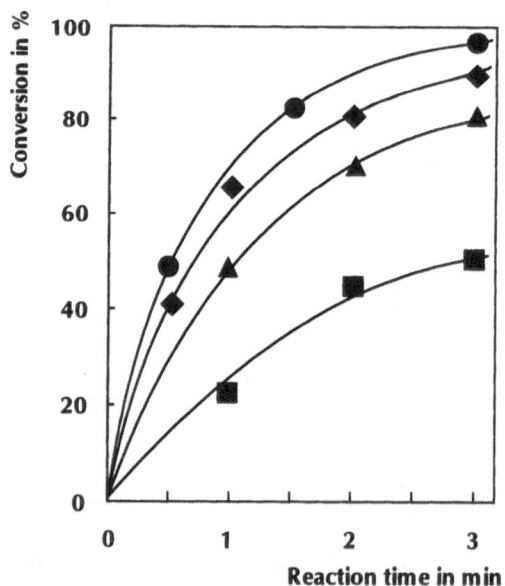

Fig. 28. Polymerization of 1,2-epoxypropane (PO) by the alcoholatealuminum porphyrin (**16**, n=10)–methylaluminum bis(2,4,6-tri-*tert*-butylphenolate) (**3f**) system, $[PO]_0/[16]_0=$ 200, $[16]_0=18.5$ mM, CH_2Cl_2 as solvent, rt, initial ratios of **3f** to PO (mol%)=0.025 (■), 0.1 (▲), 0.5 (◆), and 2.5 (●). Effect of the amount of Lewis acid **3f** on the rate of polymerization

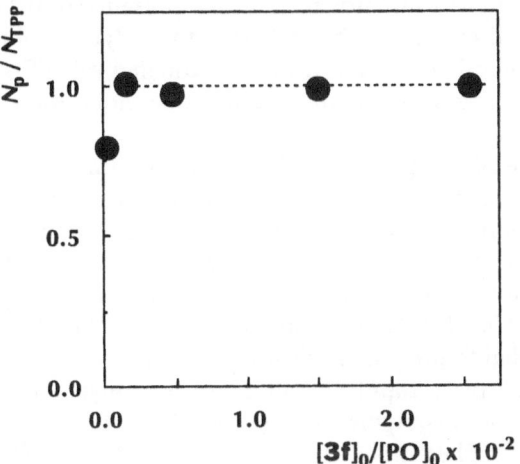

Fig. 29. Polymerization of 1,2-epoxypropane (PO) by the alcoholatealuminum porphyrin (**16**, n=10)–methylaluminum bis(2,4,6-tri-*tert*-butylphenolate) (**3f**) system, $[PO]_0/[16]_0$= 200, $[16]_0$=18.5 mM, CH_2Cl_2 as solvent, rt. Relationship between the ratio of the numbers of the molecules of polymer (N_p) to **16** (N_{TPP}) and the initial mole ratio of **3f** to PO

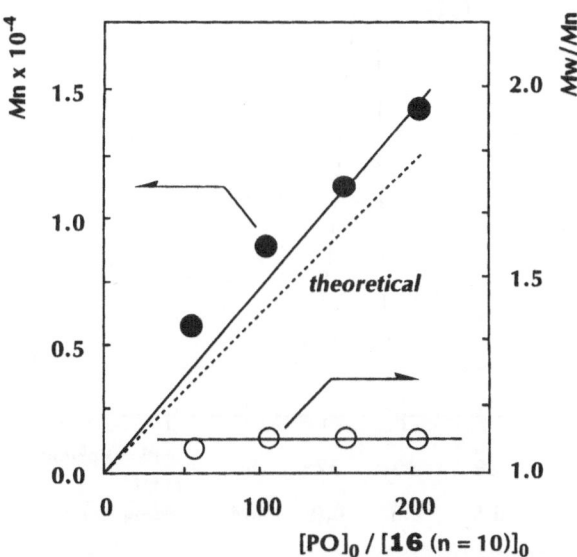

Fig. 30. Polymerization of 1,2-epoxypropane (PO) by the alcoholatealuminum porphyrin (**16**, n=10)–methylaluminum bis(2,4,6-tri-*tert*-butylphenolate) (**3f**) (1.0:0.2) system in CH_2Cl_2 at rt. Relationship between Mn (●) [Mw/Mn (○)] of the polymer formed at 100% conversion and the initial monomer-to-initiator mole ratio ($[PO]_0/[16]_0$)

changing the monomer-to-initiator ratio in the presence of 0.2 equiv with respect to **16** (n=10), where the observed Mns of the polymers obtained at 100% conversion were in good agreement with the estimated value from $[PO]_0/[16]_0$. (Fig. 30) These results indicate that the added **3f** does not initiate but only accelerates the polymerization.

The living nature of the polymerization of PO by the alcoholatealuminum porphyrin (**16**)–organoaluminum (**3f**) system was clearly demonstrated by a successful block copolymerization from PO to 1,2-epoxybutane (BO): At the first stage, polymerization of PO in CH_2Cl_2 by the **16**–**3f** system ($[PO]_0/[3f]_0/[16]_0=$ 50/1.0/0.2) was complete within 15 min at room temperature to give a polymer with Mn and Mw/Mn of 4600 and 1.04, respectively [Fig. 31 (I)]. When BO (200 equiv) was added to the polymerization mixture, the second-stage polymerization of BO took place rapidly to attain 75.2% conversion in 60 min. The GPC chromatogram of the polymer obtained at 75.2% conversion was observed to shift towards the higher molecular-weight region [Fig. 31 (II); Mn=11,300, Mw/Mn=1.18] from that of the prepolymer of PO (I), retaining the narrow MWD.

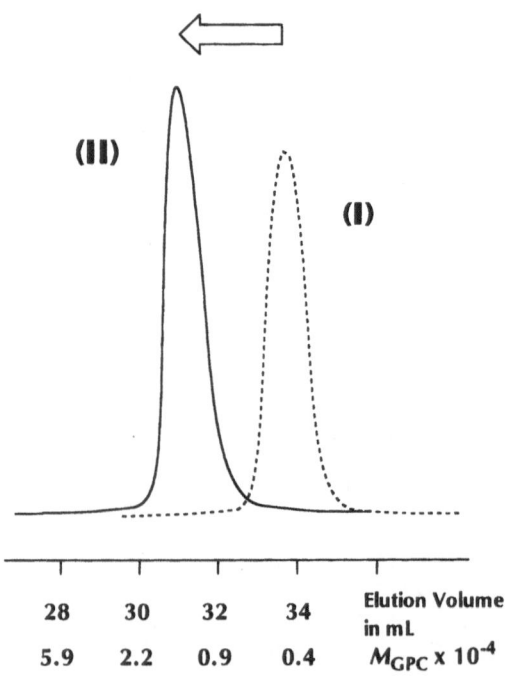

Fig. 31. Block copolymerization of 1,2-epoxypropane (PO) and 1,2-epoxybutane (BO) in CH_2Cl_2 at rt by the alcoholatealuminum porphyrin (**16**, n=10)-methylaluminum bis(2,4,6-tri-*tert*-butylphenolate) (**3f**) (1.0:0.5) system. GPC profiles of *I* The prepolymer of PO ($[PO]_0/[16]_0=50$, 100% conversion; Mn=4,600, Mw/Mn=1.04) and *II* The block copolymer of PO and BO ($[BO]_0/[16]_0=200$, 75.2% conversion; Mn=11,300, Mw/Mn=1.18)

2.6.2
Ring-Opening Polymerization of Epoxide with Various Aluminum Complexes as Initiators in the Presence of a Lewis Acid

The key point of the "high-speed living polymerization" is the steric suppression of an undesired reaction between the nucleophilic growing species and the Lewis acid, for which not only the steric bulk of the Lewis acid but also that of the porphyrin ligand is considered important. The benefit of using a Lewis acid holds even for the aluminum complexes with phthalocyanine (**11**), tetraazaannulene (**12**), and. Schiff bases (**13–15**). As initiators, these complexes exhibit much lower activity for the polymerization of PO than aluminum porphyrin **1** (X=Cl).

As reported by Spassky et al. [62], aluminum complexes of Schiff bases as initiators exhibit much lower activities than aluminum porphyrins for the ring-opening polymerization of epoxides. In fact, the polymerization of PO (500 equiv) using a Schiff base complex (Salphen)AlCl (**13**) as initiator proceeded extremely slowly at room temperature to attain only 4% conversion in 8 d. Even at 80 °C, the polymerization was slow, and required 6 d for completion, affording a polymer with broad and bimodal MWD (Fig. 32A).

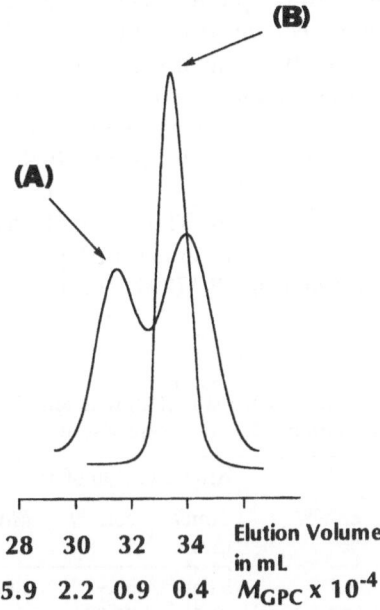

(B)

(A)

| 28 | 30 | 32 | 34 | Elution Volume in mL |
| 5.9 | 2.2 | 0.9 | 0.4 | $M_{GPC} \times 10^{-4}$ |

Fig. 32. Polymerizations of 1,2-epoxypropane (PO) by (Salphen)AlCl (**5**) and the (Salphen)AlCl (**13**)–methylaluminum bis(2,6-di-*tert*-butyl-4-methylphenolate) (**3e**) system. GPC profiles of poly(1,2-epoxypropanes) obtained with *A* **13** ([PO]$_0$/[**13**]$_0$=500, [**13**]$_0$= 28.6 mM, without solvent, 80 °C, 6 d, 100% conversion; Mn=6,700, Mw/Mn=2.31) and *B* The **13**–**3e** system ([PO]$_0$/[**3e**]$_0$/[**13**]$_0$=200/1.0/1.0, [**13**]$_0$=71.4 mM, without solvent, rt, 7 d in the absence of **3e** and 70 min after the addition of **3e**, 43.3% conversion; Mn=4200, Mw/Mn=1.18)

In the presence of methylaluminum bis(2,6-di-*tert*-butyl-4-methylphenolate) (3e), the polymerization of PO with 13 took place smoothly even at room temperature, affording a polymer with narrow MWD: An example is shown by the polymerization of PO (200 equiv) with 13 without solvent at room temperature. The polymerization proceeded up to 5.1% conversion in 7 d without 3e, while the monomer conversion reached 29.5 and 43.3% in 15 and 70 min, respectively, after the addition of 3e ($[3e]_0/[13]_0=1.0$). This corresponds to approximately 3200-times acceleration compared with the polymerization in the absence of 3e. As estimated by GPC (Fig. 32B), the Mw/Mn ratio of the polymer produced at 43.3% conversion was 1.18, and the Mn was 4200, which is close to that expected (5000) when every molecule of 13 forms one polymer molecule (Table 8, run 3). In the presence of 3e, (Salen)AlCl (14) and (Salpn)AlCl (15) were also capable of bringing about the polymerization of PO at room temperature under appropriate conditions, where the degrees of acceleration are estimated to be 1300- and 1200-times, respectively (Table 8, runs 4 and 5).

A similar Lewis acid promoted anionic chain growth was realized in the ring-opening polymerization of PO initiated with the aluminum complexes of phthalocyanine 11 and tetraazaannulene 12. Although the phthalocyanine complex (Bu$_4$Pc)AlCl (11) is structurally analogous to (TPP)AlCl (1, X=Cl), the polymerization of PO (200 equiv) with (Bu$_4$Pc)AlCl (11) alone hardly took place at room temperature, and the monomer conversion after 3 d was less than 1%. On the other hand, addition of 3e to the above mixture gave rise to the polymerization of PO, where 51.0% conversion was attained in 3 d (Table 8, run 1). The polymerization of 200 equiv of PO with a tetraazaannulene complex (Me$_4$DBTAA)AlCl (12) without solvent at room temperature proceeded to only 29.9% conversion in 4 d in the absence of 3e, but the chain growth progressed up to 74.0% conversion in only 90 s upon addition of 3e to this system. The extent of acceleration was estimated to be 5200-times (Table 8, run 2). As shown in Fig. 33, a rapid polymerization proceeded from the beginning when PO (200 equiv) was added to a mixture of 12

Table 8. Polymerization of 1,2-expoxypropane (PO) with aluminum complexes (11–15) in the presence of methylaluminum bis(2,6-di-*tert*-butyl-4-methoxyphenolate) (3e)[a]

| Run | Before addition of 3e | | | After addition of 3e | | | | |
	Initiator	Time/ days	conv[b]/ %	Time/ min	conv.[b]/ %	Mn[c]	(Mn_{calc})	Mw/Mn[c]
1	11	3	<1	3 days	50.9	1,800	(5,900)	1.14
2	12	4	29.9	1.5	73.9	5,800	(8,600)	1.40
3	13	7	5.1	70	43.3	4,200	(5,000)	1.18
4	14	7	10.9	75	44.8	2,900	(5,200)	1.06
5	15	2	2.3	30	22.1	1,800	(2,600)	1.12

[a] Without solvent under N$_2$, $[PO]_0/[initiator]_0/[3e]_0=200/1.0/1.0$, $[initiator]_0=71.4$ mM.
[b] Determined by ^{1}H NMR analysis of the reaction mixture.
[c] Estimated by GPC based on polystyrene standards.

and **3e** (1:1), where the Mn value of the produced polymer increased linearly with monomer consumption (Fig. 34), while the Mw/Mn ratio was almost constant at 1.4–1.5.

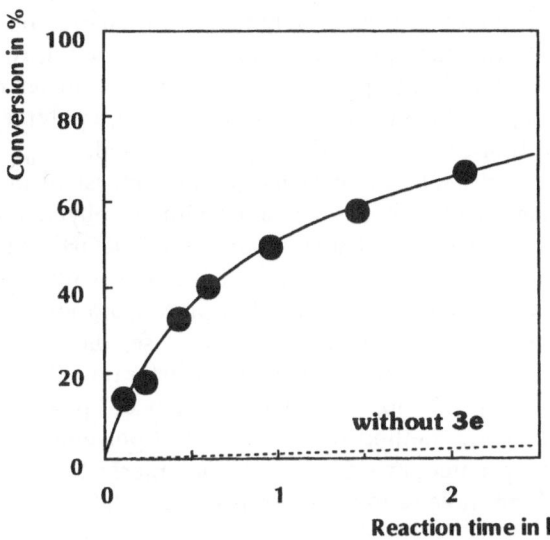

Fig. 33. Polymerization of 1,2-epoxypropane (PO) by the (Me$_4$DBTAA)AlCl (**12**)–methylaluminum bis(2,6-di-*tert*-butyl-4-methylphenolate) (**3e**) system, [PO]$_0$/[3e]$_0$/[12]$_0$=200/1.0/1.0, [12]$_0$=71.4 mM, without solvent, rt. Time-conversion curve

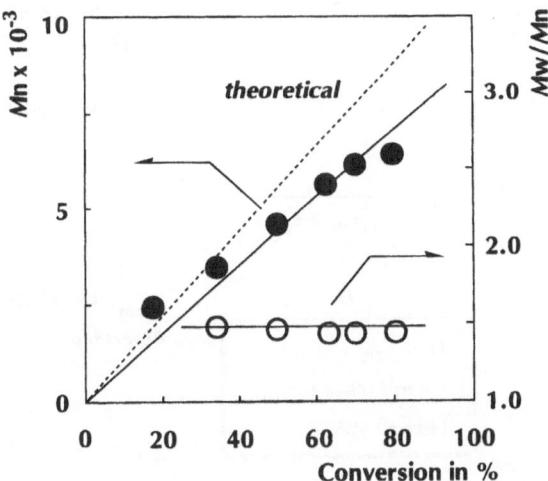

Fig. 34. Polymerization of 1,2-epoxypropane (PO) by the (Me$_4$DBTAA)AlCl (**12**)–methylaluminum bis(2,6-di-*tert*-butyl-4-methylphenolate) (**3e**) system (reaction conditions, see Fig. 33). Relationship between Mn (●) [Mw/Mn (○)] of the polymer and conversion

2.6.3
"Immortal" Polymerization of Epoxide with Aluminum Porphyrin–Alcohol Systems in the Presence of a Lewis Acid [63]

In 1985, we proposed "immortal" polymerization as a conceptually new methodology for the efficient synthesis of polymers with narrow-molecular-weight distribution [64] "Immortal" polymerization involves rapid, reversible chain transfer (exchange) reaction (Fig. 35). Since the chain transfer reaction takes place much more rapidly than the propagation reaction, a polymer with narrow MWD is formed with the number of the polymer molecules (N_p) exceeding that of the initiator molecules (N_{TPP}). This concept has been established in the ring-opening polymerization of epoxides with aluminum porphyrin initiators. In the presence of a chain transfer agent such as an alcohol (R'OH), the growing species, an alcoholatealuminum porphyrin (16_n), reacts with R'OH reversibly (Scheme 8), so that the polymerization takes place from all the molecules of 1(X=Cl) and R'OH ($N_p=N_{TPP}+N_{R'OH}$). In this case, the exchange reaction (Scheme 8) has been estimated to occur 10-times faster than the propagation reaction (Scheme 7) [64]. "Immortal" polymerization of epoxide is therefore a highly efficient synthetic method to narrow MWD polyethers. Unfortunately, however, a practical problem exists in that the polymerization slows down considerably when a high mole ratio [chain transfer agent]$_0$/[initiator]$_0$ such as 49 is applied.

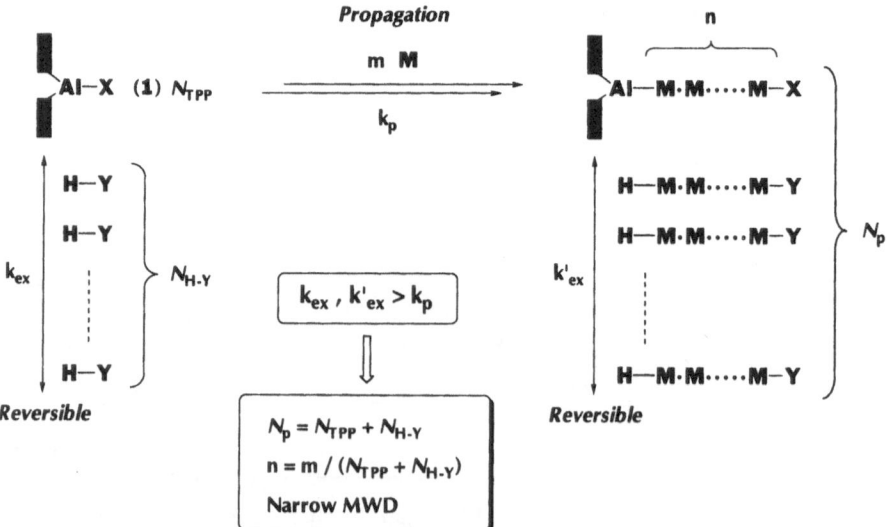

Fig. 35. Schematic representation of the concept and principle of "immortal" polymerization. *H-Y* Protic chain transfer agent; *M* monomer; k_{ex}, k'_{ex} rate constants of chain transfer (exchange) reactions; k_p rate constant of propagation reaction; N_{TPP}, N_{H-Y}, and N_p numbers of the molecules of (TPP)AlX (1), H-Y, and polymer, respectively

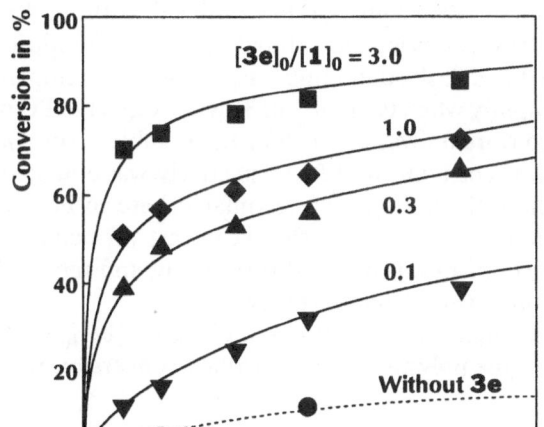

Scheme 8

Fig. 36. Polymerization of propylene oxide (PO) initiated with the (TPP)AlCl (1, X=Cl)–2-propanol (2-PrOH) system in the presence of methylaluminum bis(2,6-di-*tert*-butyl-4-methylphenolate) (**3e**) ([2-PrOH]$_0$/[PO]$_0$/[1]$_0$=9/200/1) in CH$_2$Cl$_2$ at rt. Effect of the concentration of **3e** on the rate of polymerization

This section describes the Lewis acid promoted immortal polymerization of epoxide by aluminum porphyrin–alcohol systems in the presence of an orga-noaluminum accelerator.

Polymerization of PO initiated with the (TPP)AlCl (1, X=Cl)–2-propanol (2-PrOH) system in the presence of methylaluminum bis(2,6-di-*tert*-butyl-4-methylphenolate) (**3e**) was carried out by the addition of a mixture of PO and 2-PrOH to a CH$_2$Cl$_2$ solution of a mixture of 1 (X=Cl) and **3e** at room temperature in a nitrogen atmosphere (Fig. 36). For example, when a mixture of 200 equiv of PO and 9 equiv of 2-PrOH was added to the solution of 1 containing 0.3 equiv of **3e** (0.15 mol% with respect to PO), the color of the solution immediately turned from dark reddish purple to bright reddish purple, characteristic of the alcoho-latealuminum porphyrin family. The polymerizations proceeded very rapidly with heat evolution to attain 57% monomer conversion in 3 min (▲). In contrast, the polymerization without **3e** under similar conditions proceeded rather slowly to attain only 13% conversion (●). The extent of acceleration became more explicit by increasing the amount of 3e with respect to 1 up to 3.0 (1.5 mol% with respect to PO), whereupon 70% monomer conversion was attained in only 30 s (■). GPC analysis showed that the produced polymer ($[3e]_0/[1]_0$=3.0; 100% conversion) had a narrow MWD (Mw/Mn=1.13), with a Mn value (1100) close to the theoretical one (1200) on the assumption that all the molecules of 1 and 2-PrOH had participated in initiating the polymerization (N_p/N_{TPP}=10).

Polymerization of 1000 equiv of PO initiated with 1 (X=Cl) containing 49 equiv of 2-PrOH was carried out without solvent by the addition of a mixture of PO and 2-PrOH to a flask containing 1 and 3e at room temperature. The reaction proceeded rapidly when 0.1 mol% of 3e with respect to PO was present, and 86% monomer conversion was attained in 1.5 h (Table 9, run 1), and a polymer with Mn and Mw/Mn of 1100 and 1.06, respectively, was obtained at 96% conversion (Fig. 37). The ratio N_p/N_{TPP} was estimated here to be 53, which is again close to the theoretical value of 50. In sharp contrast, the polymerization without 3e under similar conditions proceeded very slowly, and required 380 h to reach 84% monomer conversion (Table 9, run 2).

When the initial monomer-to-initiator mole ratio ($[PO]_0/[1 (X=Cl)]_0$) was increased, the Mn of the polymer formed at 100% conversion was increased line-

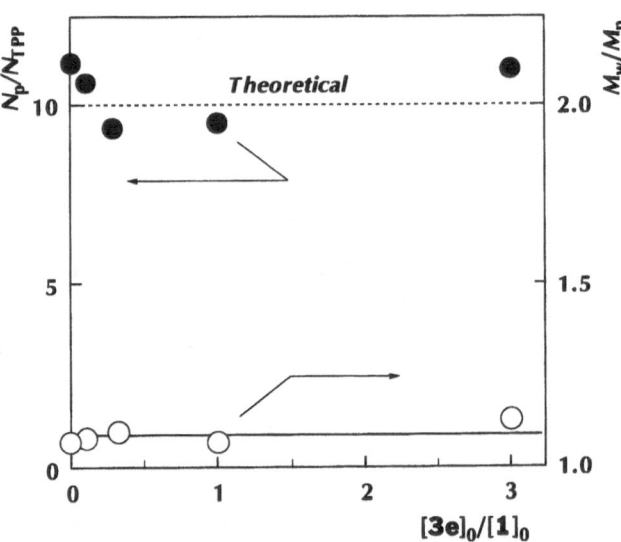

Fig. 37. Polymerization of propylene oxide (PO) initiated with the (TPP)AlCl (1, X=Cl)–2-propanol (2-PrOH) system in the presence of methylaluminum bis(2,6-di-*tert*-butyl-4-methylphenolate) (3e) ($[2$-PrOH$]_0/[PO]_0/[1]_0$=9/200/1) in CH_2Cl_2 at rt. Relationship between N_p/N_{TPP} (●) [Mw/Mn (○)] at 100% conversion and the initial mole ratio $[3e]_0/[1]_0$. N_p and N_{TPP} Numbers of the molecules of the produced polymer and initiator 1, respectively

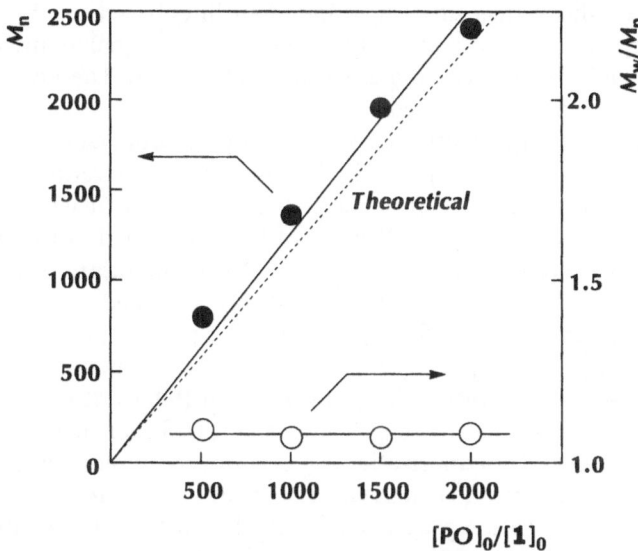

Fig. 38. Polymerization of propylene oxide (PO) initiated with the (TPP)AlCl (**1**, X=Cl)–2-propanol (2-PrOH) (1/49) system in the presence of methylaluminum bis(2,6-di-*tert*-butyl-4-methylphenolate) (**3e**) without solvent at rt. Relationship between Mn (●) [Mw/Mn (○)] of the produced polymer and the initial monomer-to-initiator mole ratio ([PO]$_0$/[**1**]$_0$) at 100% conversion

Table 9. High-speed immortal polymerization of propylene oxide (PO) with the (TPP)AlCl (**1**, X=Cl)–2-propanol (2-PrOH) system in the presence of methylaluminum bis(2,6-di-*tert*-butyl-4-methylphenolate (**3e**)[a]

Run	[2-PrOH]$_0$/[PO]$_0$/[**1**]$_0$	[**3e**]$_0$/[PO]$_0$/%	Time/h	Conv./%	Mn[b]	Mw/Mn[b]	N_p/N_{TPP}[c]
1	49/1000/1	0.1	1.5	86	900	1.10	55
2	49/1000/1	0	380	84	1100	1.08	45
3	99/2000/1	2.0	9	80	800	1.10	120
4	299/6000/1	3.3	19	81	800	1.11	340
5	999/20,000/1	1.5	250	62	600	1.13	1150
6d	999/20,000/1	1.5	120	61	600	1.14	1100
7	999/20,000/1	0	1900	10	–	–	–

[a] Without solvent under N_2 at rt.
[b] Estimated by GPC based on poly(propylene glycol) standards.
[c] Number of polymer molecules (N_p)/number of molecules of
   1 (N_{TPP})=58.08 ([PO]$_0$/[**1**]$_0$)×(conversion/100)×M_n^{-1}.
[d] At 60 °C

arly along the theoretical line (Fig. 38, broken line),based on the assumption that the number of the produced polymer molecules is equal to the sum of those of the molecules of 1 (X=Cl) and 2-PrOH (N_p/N_{TPP}=50). The Mw/Mn ratios of the produced polymers remained almost constant at 1.1.

By taking advantage of this, narrow MWD poly(propylene oxide)s with an N_p/N_{TPP} of 55 to 1100 were synthesized by increasing the ratio of 2-PrOH to 1 (X=Cl) from 49 to 1000 in the presence of 0.1–3.3 mol% of 3e with respect to the monomer. For example, when 2000 equiv of PO were added to 1 in the presence of 99 equiv of 2-PrOH without solvent at room temperature, a polymer with Mw/Mn of 1.10 and N_p/N_{TPP} of 120 (Mn=800) was formed in 9 h (80% conversion) when 2.0 mol% of 3e with respect to PO was present (Table 9, run 3). Similarly, at a ratio [2-PrOH]$_0$/[PO]$_0$/[1]$_0$ of 299/6000/1, a polymer with Mw/Mn of 1.11 and N_p/N_{TPP} of 340 (Mn=800) was formed in 19 h (81% conversion) when 3.3 mol% of 3e with respect to PO was present (Table 9, run 4). Finally, we attempted the immortal polymerization of PO with 1 by employing 999 equiv of 2-PrOH and 20,000 equiv of PO. In the absence of 3e at room temperature without solvent, the polymerization proceeded extremely slowly to attain only 10% monomer conversion even in 1900 h (run 7). On the other hand, when 1.5 mol% of 3e with respect to PO was present under similar conditions, 62% monomer conversion was attained in 250 h, affording a polymer with Mw/Mn of 1.13 and N_p/N_{TPP} of 1150 (Mn=600) (Table 9, run 5).

In "immortal" polymerization, a reversible chain transfer reaction takes place much more rapidly than a propagation reaction, thereby uniformity of the molecular weight of polymer is realized. The successful high-speed "immortal" polymerization assisted by a Lewis acid, mentioned above, suggests that not only the propagation step (Scheme 7) but also the chain transfer process (Scheme 8) is accelerated by a Lewis acid.

In order to examine the effect of the Lewis acid on the rate of the chain transfer reaction, the ^1H NMR saturation transfer method [65] was applied to the model system, a mixture of (2-propanolate)aluminum porphyrin [(TPP) AlOCH(CH$_3$)$_2$, 1, X=O^{i}Pr] and 2-PrOH (1:13) in theabsence (Fig. 39) and presence (Fig. 40) of the Lewis acid 3e (3 equiv) in C$_6$D$_6$ at 21 °C.

In Fig. 39A, the signals due to the CH$_3$ groups of 1 (X=O^{i}Pr) and 2-PrOH are observed at δ –1.40 (a) and 1.20 (b), respectively. Although the signal b was irradiated to saturate the CH$_3$ protons of 2-PrOH, the signal a was only slightly decreased in intensity (Fig. 39B), where the relative intensity with and without irradiation (I_{irrad}/I_0) was approximately 0.96. In case where the Lewis acid 3e is present, one important thing to note is that the alcoholatealuminum porphyrin (1, X=O^{i}Pr) can survive without degradative nucleophilic attack to the Lewis acid, as evidenced by the fact that the relative intensity of the signal a (6H) to that of the porphyrin pyrrole-β-protons (δ 9.33 ppm, 8H) satisfied the required ratio. More interestingly, upon irradiation of the signal b, the intensity of the signal a fell to 31% of the original intensity (I_{irrad}/I_0=0.31) (Fig. 40). Therefore, it is concluded that the alcoholate–alcohol exchange reaction at the axial position of the aluminum porphyrin (Scheme 8) is accelerated by the Lewis acid. Here, the ratio I_{irrad}/I_0 is given by the following equation:

$$I_{irrad}/I_0 = 1/(1 + k_{ex} \infty T_1)$$

where k_{ex} and T_1 represent the rate constants of the exchange reaction and the spin lattice relaxation time of the CH_3 protons of 1 ($X = O^iPr$), respectively.

Thus, from the above two I_{irrad}/I_0 ratios observed in the absence and presence of 3e, the exchange reaction is approximately 50-times accelerated by 3e under the experimental conditions. In the presence of 3e, the alcohol is considered to turn more acidic upon possible coordination to the Lewis acid, thereby the exchange reaction with 1 ($X = O^iPr$) is accelerated.

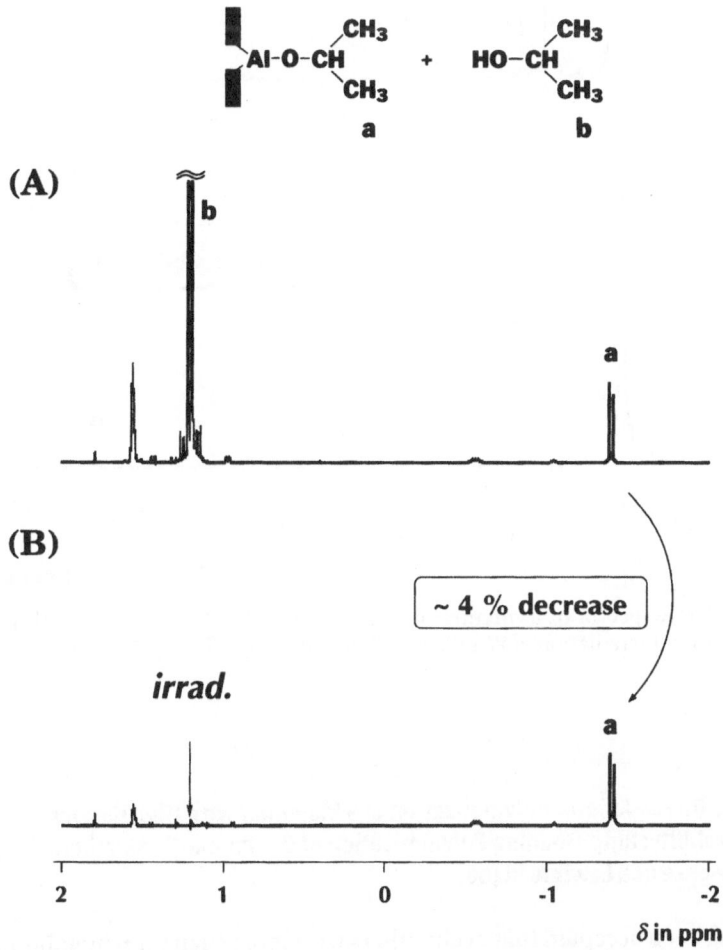

Fig. 39. ^{1}H NMR spectra of a mixture of 1 ($X = O^iPr$) and 2-PrOH (1/13) in C_6D_6 at 21 °C. *A* Without irradiation. *B* With irradiation of the signal b ([1] = 25 mM)

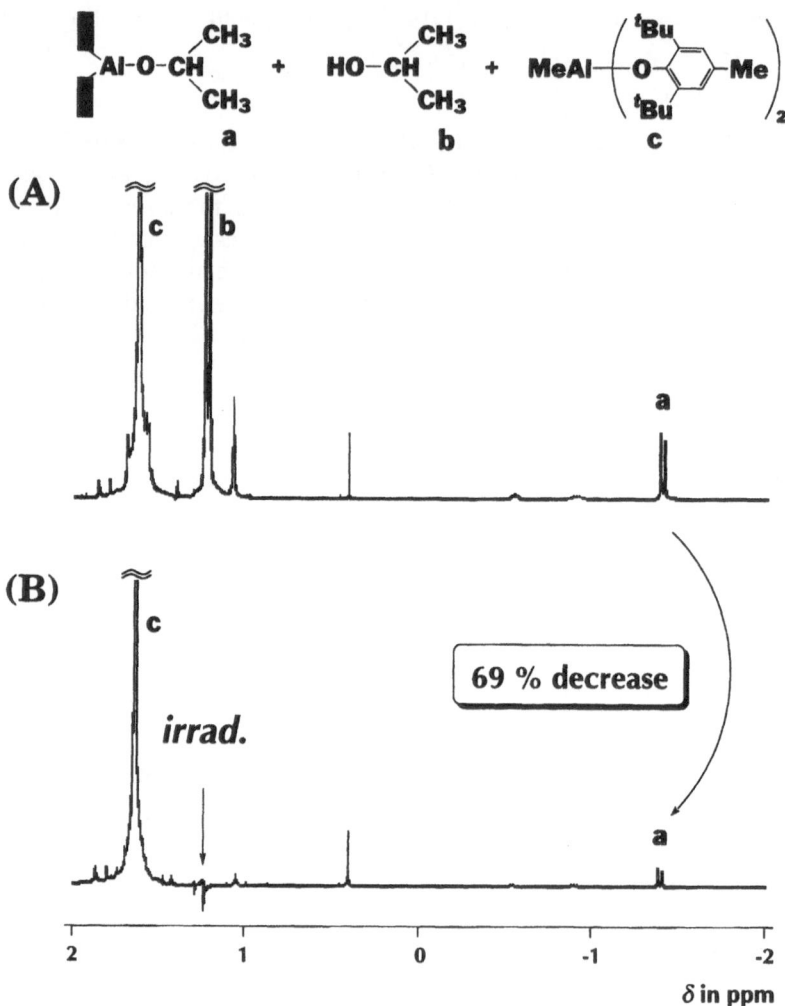

Fig. 40. ^{1}H NMR spectra of a mixture of **1** (X=O^{i}Pr), 2-PrOH, and **3e** (1/13/3) in C$_6$D$_6$ at 21 °C: *A* Without irradiation. *B* With irradiation of the signal b ([**1**]=25 mM)

2.6.4
Lewis Acid Driven Anionic Polymerization of a Monomer with High Cationic Polymerizability: Ring-Opening Polymerization of Oxetane with Aluminum Porphyrin in the Presence of a Lewis Acid [66]

It has long been accepted that cyclic ethers with more than four-membered rings are, in contrast to epoxides, only cationically polymerizable due to the high basicities of their ether oxygen atoms. The cationic polymerization involves *O*-

alkylation of the monomer to give a highly reactive cyclic oxonium ion as the growing species, which is regarded as the monomer activated for nucleophilic attack. From this standpoint, one can expect that even cyclic ethers with high basicities, upon activation by electrophiles, can undergo anionic ring-opening polymerization.

As described in the previous sections, the living anionic polymerizations of epoxides and methacrylic esters initiated with aluminum porphyrins 1 [67] are dramatically accelerated upon addition of sterically hindered Lewis acids such as 3 [68,69], where the monomers are coordinated to and activated for nucleophilic attack by the Lewis acids. Successful extension of this method is the living anionic polymerization of oxetane [70].

Although 1 (X=Cl) is an excellent initiator for the living anionic polymerization of epoxides, no polymerization of oxetane took place throughout the observation over a period of 3 h. On the other hand, when 3e (6 equiv with respect to 1) [71] was added to the reaction mixture ([oxetane]$_0$/[1]$_0$=100), the polymerization did take place to attain 47.4 and 88.4% conversion in 0.5 and 3 h, respectively. The produced polymer showed a sharp GPC chromatogram, and the Mn value [72] was increased proportionally to monomer conversion, retaining the narrow MWD (Mw/Mn~1.1) (Fig. 41). The degree of polymerization of the polymer was in good agreement with the mole ratio of the monomer reacted to the initiator, indicating that every initiator molecule produces one polymer molecule. When 3b was used as the Lewis acid, a polymer with Mn and Mw/Mn of 19,000 (Mn_{theory}=23,000) and 1.11, respectively, was obtained at 100% conversion in 13.5 h by applying the initial monomer-to-initiator mole ratio of 400.

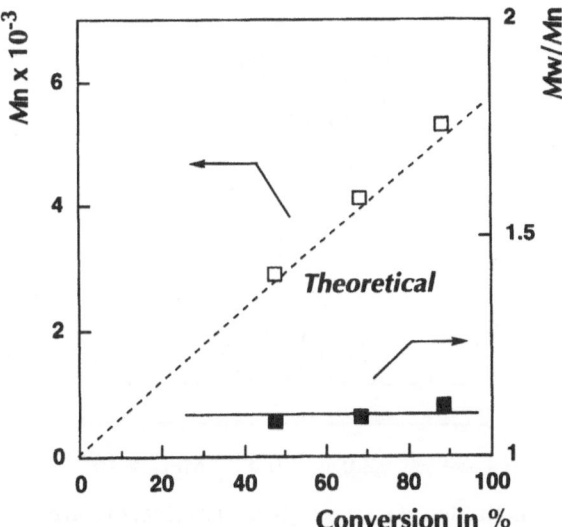

Fig. 41. Polymerization of oxetane with the 1 (X=Cl)–3e system in CH$_2$Cl$_2$ at rt (~20 °C) ([oxetane]$_0$/[1]$_0$/[3e]$_0$=100/1/6). Mn (Mw/Mn)–conversion relationship

After the completion of the first-stage polymerization of oxetane ([oxetane]$_0$/[1 (X=Cl)]$_0$/[3e]$_0$=50/1/3) at room temperature, 150 equiv of oxetane with respect to 1 were newly added to the system, whereupon the second-stage polymerization ensued, and the Mn of the polymer was increased from 2900 (Mw/Mn=1.09) to 8300 (Mw/Mn=1.22), retaining the narrow MWD.

A preliminary ^{1}H NMR study indicated that 1 (X=Cl) is able to react with oxetane even in the absence of 3e at room temperature affording, however, a one-to-one adduct, (TPP)AlOCH$_2$CH$_2$CH$_2$Cl [δ –1.43 (t), –1.26 (m), 0.82 (t), CD$_2$Cl$_2$], without any polymeric products. In conformity with this observation, the polymerization of oxetane (100 equiv) at room temperature using a (porphinato)aluminum alcoholate such as 1 (X=O^{i}Pr) as initiator also took place in the presence of 3e (3 mol% with respect to oxetane), affording at 100% conversion a polymer with Mn and Mw/Mn of 8900 and 1.24, respectively. ^{1}H NMR analysis of the polymer, after hydrolytic workup, showed the presence of a 2-propoxy end group [δ 1.15, CH$_3$ (d)] originating from 1 (X=O^{i}Pr) but no phenoxy group from 3e.

By using the aluminum porphyrin–Lewis acid system, we attempted the synthesis of a narrow MWD block copolymer from oxetane and methyl methacrylate (MMA). Methacrylic monomers can be polymerized radically and anionically but not cationically, so a block copolymer of oxetane and methyl methacrylate has never been synthesized. As already reported, methacrylic monomers undergo accelerated living anionic polymerization with the (TPP)AlMe (1, X= Me)–3e system via a (porphinato)aluminum enolate as the growing species.

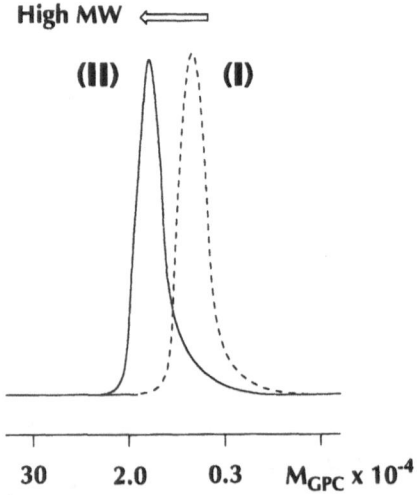

Fig. 42. Block copolymerization of methyl methacrylate (MMA) and oxetane with the 1 (X= Me)–3e system ([MMA]$_0$/[oxetane]$_0$/[1]$_0$/[3e]$_0$=100/50/1/6). GPC chromatograms of the prepolymer of MMA *I* Mn=6500, Mw/Mn=1.13 and the block copolymer of MMA and oxetane and *II* Mn=15,000, Mw/Mn=1.20

Thus, MMA (50 equiv with respect to 1) was first polymerized with 1 in the presence of 3e (6 equiv) {PMMA at 100% conversion, Mn=6500 [Mn_{theory}=5000], Mw/Mn=1.13 [Fig. 42 (I)]}, and then oxetane (150 equiv) was added to this system, whereupon the second-stage polymerization ensued and was complete within 14 h. The GPC chromatogram of the product [Fig. 42 (II)] clearly shifted towards the higher molecular-weight region, while retaining the narrow MWD (Mn=15,000 [Mn_{theory}=11,000], Mw/Mn=1.20). The successful formation of the PMMA–polyoxetane block copolymer demonstrates that the polymerization of oxetane initiated with 1 (X=Cl) in the presence of 3e is *anionic*.

2.6.5
Ring-Opening Polymerization of Lactone with Aluminum Porphyrin in the Presence of a Lewis Acid [73]

Polymerizations of six- and four-membered lactones initiated with aluminum porphyrins [(TPP)AlX, 1] proceed via (porphinato)aluminum alkoxide and carboxylate as the growing species, respectively (Schemes 9 and 10) [74,75] to give polyesters with narrow MWD, although it is well known that the polymerization

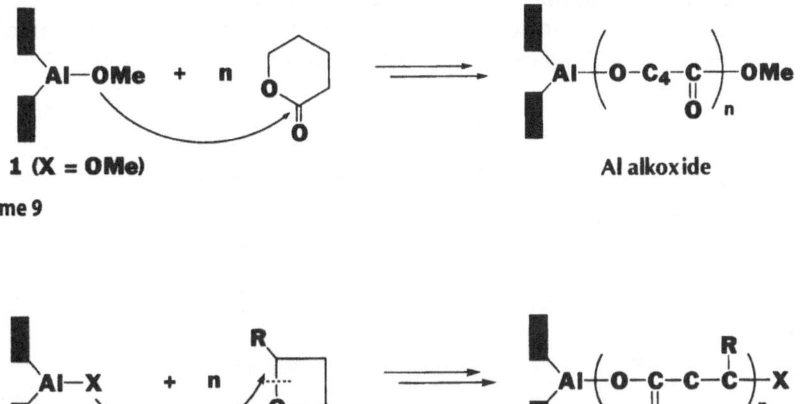

1 (X = OMe)   Al alkoxide

Scheme 9

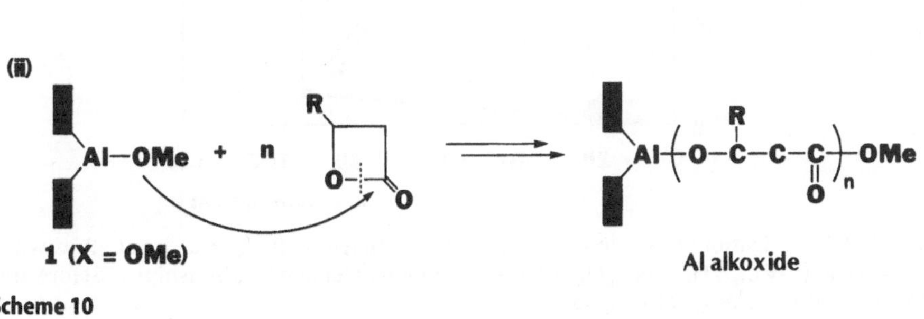

(i)

1 (X = Cl , O$_2$CR)   Al carboxylate

(ii)

1 (X = OMe)   Al alkoxide

Scheme 10

of lactone initiated with boron trifluoride, ferric chloride, aluminum chloride, or aluminum isopropoxide cannot avoid inter- or intramolecular transesterification to cause degradation of the once formed polyester by the attack of the active chain ends to the polymer ester groups [76].

In the present study, the novel concept of the "Lewis acid assisted living polymerization with the aluminum porphyrin–methylaluminum diphenolate (3) system" was successfully extended from the accelerated living addition polymerization of alkyl methacrylates to the accelerated living ring-opening polymerizations of lactones.

(I) Polymerization of δ-Valerolactone with Aluminum Porphyrin in the Presence of an Organoaluminum Compound. Aluminum porphyrin brings about the living ring-opening polymerization of δ-valerolactone (δ-VL) to give polymers with narrow MWD [75]. A representative initiator is (TPP)AlOMe (1, X=OMe), and the growing species is an aluminum alkoxide complex of porphyrin (Scheme 9). However, the polymerization of δ-VL with 1 (X=OMe) ($[\delta\text{-VL}]_0/[1\ (\text{X=OMe})]_0=200$) at 50 °C without solvent proceeded rather slowly to attain only 7% conversion in 91 h (Fig. 43a). On the other hand, when 3 equiv of 3e with respect to 1 was added at 50 °C to this reaction mixture, a remarkable rise in the viscosity of the polymerization system was observed to attain 89% conversion in only 15 min after the addition of 3e. This corresponds to 2100-times acceleration of polymerization. The M_n increased proportionally to the conversion of the monomer. The degree of polymerization of the polymer was in good agreement with the mole ratio of the

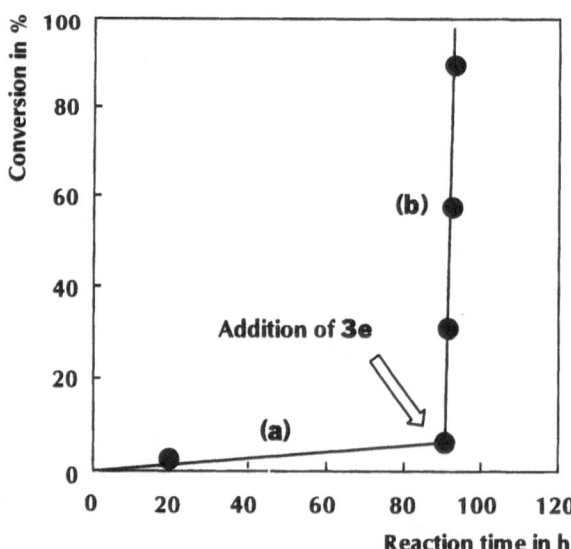

Fig. 43. Polymerization of δ-valerolactone (δ-VL) initiated with 1 (X=OMe) at 50 °C without solvent, $[\delta\text{-VL}]_0/[1]_0=200$, $[1]_0=42.4$ mM. Time-conversion relationships *a* Before and *b* After addition of 3e ($[3e]_0/[1]_0=3$)

monomer reacted to the initiator, indicating that every initiator molecule produces one polymer molecule. The MWD of the polymer was very narrow at the earlier stage of the polymerization (Mw/Mn=1.05 [Mn=2200] at 7% monomer conversion), but tended to become broader at the later stage (Mw/Mn=1.58 [Mn=27,000] at 89% conversion). This broadening of the MWD is attributed to the possible activation of the ester carbonyl group of the formed polymer by **3e**, which would lead to the accelerated intra- and/or intermolecular attack of the growing species to the ester groups in the polymer main chain. In order to examine this possibility, the polymerization of δ-VL (50 equiv) initiated with **1** (X=OMe) in CH_2Cl_2 at room temperature was brought to 89% conversion in 330 h to give the polymer with Mw/Mn of 1.20 (Mn=5400). Then 3 equiv of **3e** was added to the mixture, and it was left to stand for 50 h, whereupon the MWD became much broader (Mw/Mn= 1.50 [Mn=4200]).

Methylaluminum diphenolates (**3f,g** and **j-m**) brought about the accelerated polymerization of δ-VL in dichloroethane at 50 °C, corresponding to an acceleration by a factor of 100–4800 (**3f,g** and **j-m**) compared with the rate of polymerization in the absence of **3** under similar conditions (Table 10, run 8), where the Mn of the produced polymer was about 25,000 at 65–75% conversion (**3f,g** and **j-l**) except for the case using **3m**. On the other hand, a remarkable difference in the MWD of the produced polymers at about 70% conversion was observed. In the case when **3f** was used , the MWD became broader at the later stage [Fig. 44 (□)] to give the polymer with the Mw/Mn of 1.32 at 68% conversion (Table 10, run 1). A similar phenomenon was observed in the polymerization in the presence of **3m**. However, the polymerizations in the presence of **3g**, and **j-l** gave the polymers with narrow MWD (Mw/Mn=1.1–1.2) at 65%–75% conversion, respectively (Table 10, runs 2–5). In particular, in the case when using **3l** (Table 10, run 5), the MWD was kept very narrow throughout the polymerization (Mw/Mn=1.12–1.14) [Fig. 44 (■)].

Table 10. Polymerization of δ-VL with (TPP)AlOMe (**1**, X=OMe) in the presence of various methylaluminum diphenolates **3** and methylaluminum bis(triphenylcarbinolate) (**4b**)[a]

Run	3 or 4b	Time/h	Conversion/%[b]	Mn[c]	Mw/Mn[c]
1[d]	3f	1.0	67.9	26,000	1.32
2	3g	3.5	70.1	21,000	1.19
3	3j	4.8	64.7	25,000	1.21
4[e]	3k	9.0	73.7	19,000	1.17
5	3l	3.5	74.0	24,000	1.13
6	3m	5.2	66.2	28,000	1.47
7	4b	63.0	83.3	15,000	1.50
8	f	147	3.6	–	–

[a] In DCE as solvent at 50 °C, [δ-VL]$_0$/[**3** or **4b**]$_0$/[**1**]$_0$=200/3/1, [**1**]$_0$=14.6 mM.
[b] Determined by ^{1}H NMR.
[c] Estimated by GPC based on polystyrene standards.
[d] [**1**]$_0$=15.7 mM.
[e] [**1**]$_0$=11.3 mM.
[f] Without **3** or **4b**.

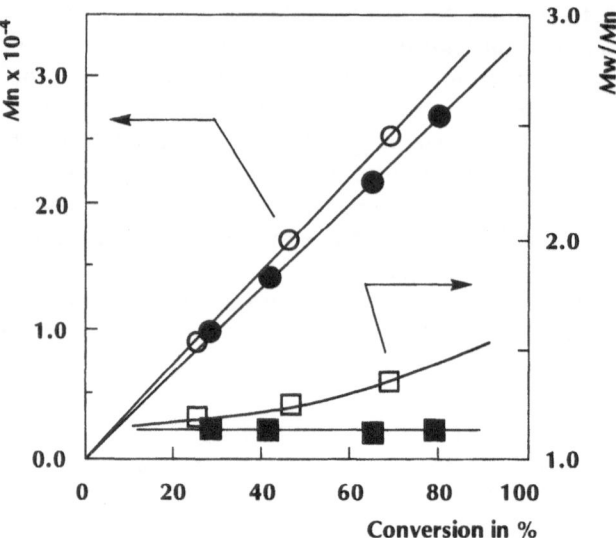

Fig. 44. Polymerization of δ-valerolactone (δ-VL) initiated with 1 (X=OMe) in the presence of **3f** and **3l** at 50 °C in DCE, $[δ\text{-VL}]_0/[3]_0/[1]_0=200/3/1$, $[1]_0=15.7$ mM (**3f**), 14.6 mM (**3l**). Relationships between the monomer conversion and Mn (**3f** ○, **3l** ●) [Mw/Mn (**3f** □, **3l** ■)]

Coordinative interaction between methylaluminum diphenolates (**3f** and **l**) and δ-VL or poly(δ-VL) was investigated by ^{13}C NMR in CD_2Cl_2 at 23 °C. The spectrum of δ-VL alone is shown in Fig. 45 (I), where the signals a (δ 170.7 ppm), b (δ 69.1 ppm), c, d, and e (δ 29.5, 22.0, and 18.8 ppm) were due to C=O, CH_2O, and other CH_2 groups, respectively. When the spectrum of the mixture at the molar ratio of **3f**/δ-VL of 1.5:1 [Fig. 45 (II)] was compared with that of δ-VL alone [Fig. 45 (I)], each signal for δ-VL in the mixture shifted to the signals a' (δ 181.5 ppm), b' (δ 74.3 ppm), c', d', and e' (δ 29.9, 21.1, and 17.2 ppm), respectively. The extent of the shift was most remarkable for the signal due to the carbonyl carbon (Δδ 10.8 ppm). Similar shifts were observed in the case of the mixture of **3l** and δ-VL ($[3l]_0/[δ\text{-VL}]_0=1.5$), where the signal assigned to the C=O shifted most remarkably (downfield shift of 10.1 ppm) [Fig. 45 (III)]. Figur 46 shows the ^{13}C NMR spectra of poly(δ-VL) alone (I), poly(δ-VL) in the mixtures of **3f** and poly(δ-VL) (II), and **3l** and poly(δ-VL) (III) ($[3]_0/[$monomer units in poly(δ-VL)$]_0=1.0$) [77]. When the spectrum of the mixture was compared with that of poly(δ-VL) alone, each signal due to poly(δ-VL) in the mixture turned out to shift similarly to δ-VL in the presence of **3**, and the extent of the shift was most noticeable in the signal assigned to the carbonyl carbon, where the signal A (δ 173.6 ppm) shifted to the signal A' (δ 182.8 ppm) and A" (δ 183.6 ppm) in the presence of **3f** and **3l** (downfield shifts of 9.2 and 10.0 ppm), respectively. However, a clear difference was observed in these two spectra. In the spectrum of the mixture of **3f** and poly(δ-VL) [Fig. 46 (II)], no signals were observed at the same chemical shift as those due to poly(δ-VL) in the absence of **3f** [Fig. 46 (I)]. On the

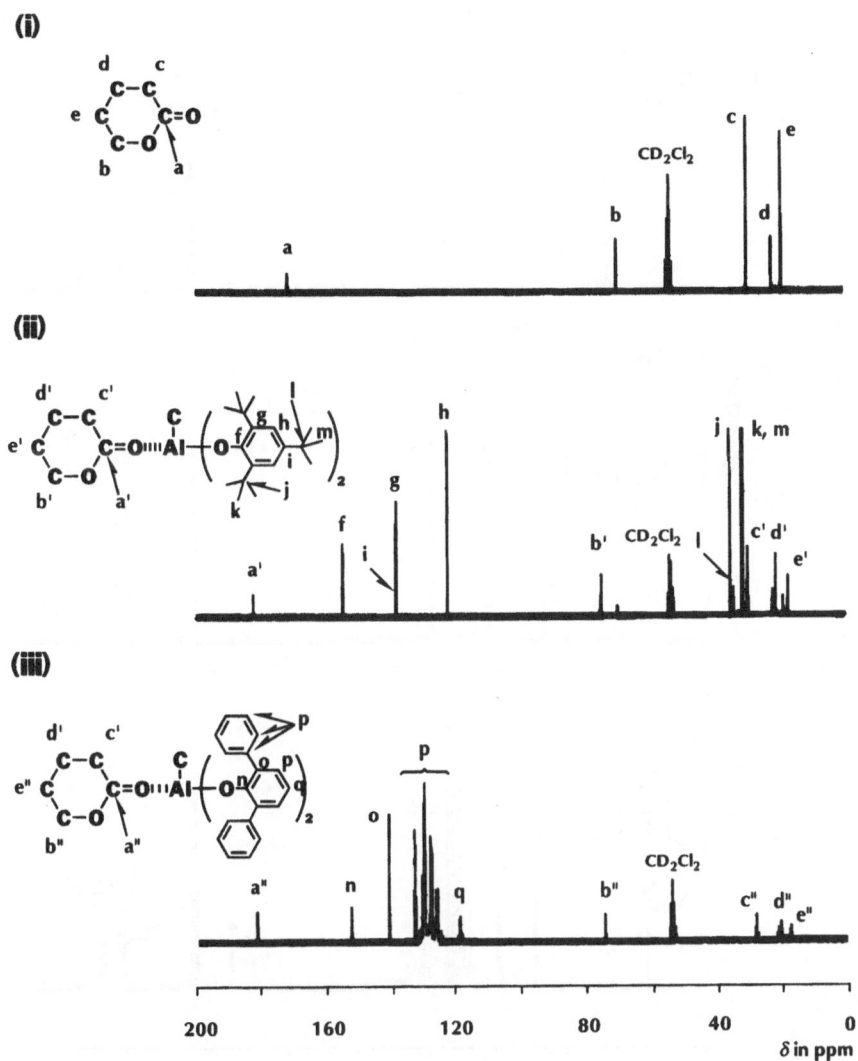

Fig. 45. ^{13}C NMR spectra in CD_2Cl_2 at 23 °C of I δ-Valerolactone (δ-VL) ([δ-VL]$_0$=0.3 M), II A mixture of **3f** and δ-VL, and III A mixture of **3l** and δ-VL ([δ-VL]$_0$/[**3**]$_0$=1.5, [δ-VL]$_0$= 0.3 M

other hand, the spectrum of the mixture of **2f** and poly(δ-VL) [Fig. 46 (III)] showed the signals at the original chemical shifts [Fig. 46 (I)] in addition to the shifted signals, where the relative intensity ratio of the shifted signals to the original signals was 4:6. These observations indicate that **3l** interacts more strongly with δ-VL than with poly(δ-VL), while **3f** possibly interacts almost equally with δ-VL and poly(δ-VL).

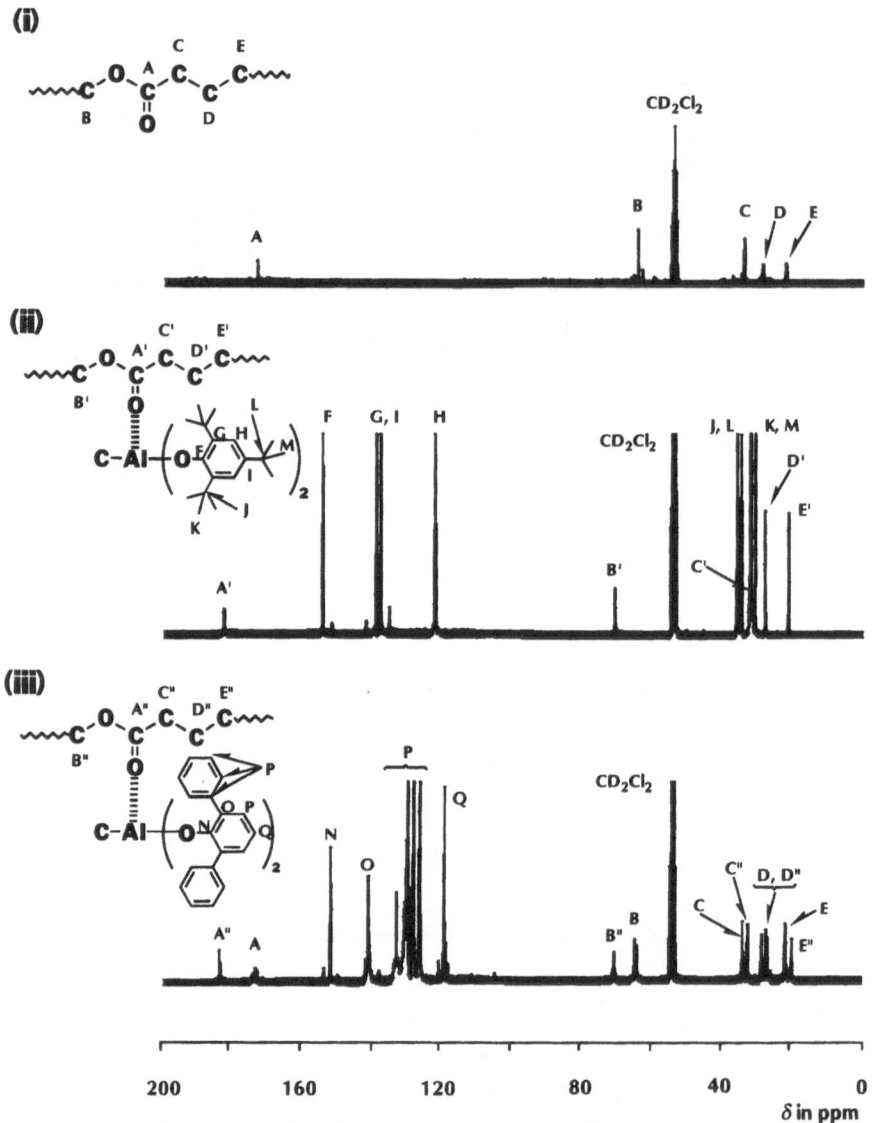

Fig. 46. ^{13}C NMR spectra in CD$_2$Cl$_2$ at 23 °C of *I* Poly(δ-valerolactone) [poly(δ-VL)] ([monomer units in poly(δ-VL)]$_0$=0.6 M), *II* A mixture of **3f** and poly(δ-VL), and *III* A mixture of **3l** and poly(δ-VL) ([monomer units in poly(δ-VL)]$_0$/[**3**]$_0$=1.0, [monomer units in poly(δ-VL)]$_0$=0.3 M)

As described above, methylaluminum phenolates (**3e–m**) accelerate the polymerization of δ-VL initiated with **1** (X=OMe) (Fig. 43 and Table 10). Together with the results of the NMR observations, the accelerated polymerization of δ-VL with the **1** (X=OMe)–**3** system is concluded to be the result of (1) coordinative activation of δ-VL by **3** for the nucleophilic attack of **1** (X=O-polymer), and (2) suppression of the undesired degradative attack of **1** to the Lewis acidic center of **3** because of the steric repulsion between the bulky porphyrin ligand of **1** and the substituents at the *ortho* positions of the phenolate ligands of **3**.

The substituents on the phenolate ligand of **3** are considered to affect the ability of promoting the possible side reaction in the later stage of the polymerization of δ-VL such as transesterification. The NMR analysis of the mixture of **3f** and poly(δ-VL) (Fig. 46) indicate a coordinative interaction between **3f** and poly(δ-VL) similar to that between **3f** and δ-VL. This coordinative interaction between **3f** and poly(δ-VL) probably promotes the attack of the growing species to the ester groups in the produced polymer main chain (transesterification) proceeding in competition with the accelerated propagation, and thus results in the broadening of the MWD at the later stage of the polymerization. On the other hand, **3l** coordinates selectively with the δ-VL monomer and thus effectively accelerates the polymerization of δ-VL; however, the interaction of **3l** with poly(δ-VL) is not enough to promote the unfavored transesterification [Fig. 46 (III)], where the MWD of the produced polymer is kept very narrow throughout the polymerization.

(II) Polymerization of β-Butyrolactone with Aluminum Porphyrin in the Presence of an Organoaluminum Compound. β-Lactone has been known to undergo living polymerization initiated with aluminum porphyrins such as (TPP)AlCl (**1**, X=Cl) and (TPP)AlO$_2$CR (**1**, X=O$_2$CR) [74]. The reaction proceeds via the cleavage of the CO$_2$–C bond, and the growing species is an aluminum carboxylate [Scheme 10 (I)]. On the other hand, we have revealed more recently that the polymerization of β-lactone also takes place with (TPP)AlOR (**1**, X=OR) as initiator though more slowly, where the ring is cleaved at the O–C=O bond, the growing species being an aluminum alkoxide, as evidenced by the ^{1}H NMR studies of the reaction mixture [Scheme 10 (II)]. The ^{1}H NMR spectrum in CDCl$_3$ for the reaction mixture of β-BL and **1** (X=Cl) (10:1) for 200 h showed signals due to (TPP)Al-O-CH(CH$_3$), (TPP)Al-O-CH(CH$_3$)-, and (TPP)Al-O-CH(CH$_3$)-CH$_2$- at δ –1.95 (d, 3H), –1.84 (m, 1H), and –1.21 and –0.39 (dq, 2H) ppm, respectively [74a], while the signals due to (TPP)Al-O-C(=O)-CH$_2$-CH(CH$_3$)- and (TPP)Al-O-C(=O)-CH$_2$-CH(CH$_3$)- overlapping at δ –0.4–0.6 ppm (5H) [74a] were not observed.

A methylaluminum diphenolate such as **3e** also exhibits an accelerating effect on the polymerization of β-lactone, and the extent of acceleration depends on the mode of the ring scission. For example, in the polymerization of β-butyrolactone (β-BL, 300 equiv) initiated with **1** (X=Cl) in CH$_2$Cl$_2$ at room temperature, the addition of **3e** (1 equiv) brought about a 2-times acceleration (Fig. 47, ■→□). The Mn (5100 [Mw/Mn=1.1] at 22% conversion; 288 h) of the polymer formed in the absence of **3e** and that (15,600 [Mw/Mn=1.2] at 57% conversion; 384 h) of the polymer obtained in the presence of **3e** were close to the expected values (5700 and 14,700, respectively) from the initial β-BL-to-**1** (X=Cl) mole ratio and the monomer conversion by assuming that every molecule of **1** produced one polymer molecule. In contrast, the polymerization initiated with

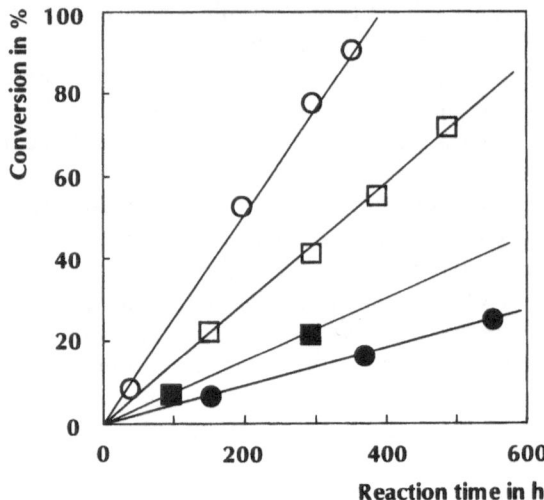

Fig. 47. Polymerizations of β-butyrolactone (β-BL) initiated with 1 (X=Cl) in the absence (■) and presence (□) of **3e**, and with 1 (X=OMe) in the absence (●) and presence (○) of **3e** at rt in CH_2Cl_2, $[\beta\text{-BL}]_0/[1]_0=300$, $[1]_0=29.4$ mM, $[3e]_0/[1]_0=0$ (■ and ●) or 3 (□ and ○)

(TPP)AlOMe was slower (26% conversion in 552 h; $Mn=6100$ [$Mw/Mn=1.2$]) than by 1 (X=Cl), but the accelerating effect of **3e** was more remarkable (6-times) for the 1 (X=OMe)-initiated polymerization (86% conversion in 362 h; $Mn=19{,}500$ [$Mw/Mn=1.2$]) (Fig. 47, ●→○).

The acceleration by **3e** is also considered to be due to the coordination of the carbonyl group of β-BL to **3e**, as indicated by the broadening of the $C=O$ signal (δ 167.7 ppm) of β-BL in the ^{13}C NMR spectrum when mixed with **3e** (1:1) in CD_2Cl_2. The coordination of the monomer is considered to affect the reactivity directly when the carbonyl group is attacked by the growing species, while the attack at the β-carbon remote from the carbonyl group is less affected by the co-ordination, resulting in different extents of acceleration depending on the mode of ring cleavage.

Irrespective of the structure of the initiator employed (1, X=OMe, Cl), the MWD of the poly(β-BL) remained narrow throughout the accelerated polymerization in the presence of **3e** ($Mw/Mn=1.1$–1.2), in sharp contrast to the broadened MWD of the poly(δ-VL) at the later stage of the polymerization with 1 (X=OMe) in the presence of **3e** ($Mw/Mn\sim1.5$).

3
Polymerization with Zinc Porphyrin

3.1
Living and Immortal Polymerizations of Episulfide Initiated with Zinc Porphyrin [79]

Aluminum porphyrins are good initiators for the living addition polymerizations of methacrylic ester, acrylic ester, and methacrylonitrile, and the ring-opening polymerizations of epoxide, lactone, and lactide. However, contrary to expectation, by the ring-opening polymerization of 1,2-epitiopropane (propylene sulfide, PS), a three-membered cyclic thioether, initiated with aluminum porphyrin, a polymer of a narrow MWD was not obtained under similar conditions. On the other hand, zinc complexes of N-substituted porphyrins are excellent initiators for the living polymerization of PS which proceeds at the zinc–axial group bond of the initiator (Scheme 11).

Among the zinc N-methylporphyrins having various axial groups, such as SPr, Cl, OAc, only the SPr complex initiated the polymerization of PS to give a polythioether with a narrow MWD. The polymerization of PS with (NMTPP)ZnSPr (17, X=SPr) ($[PS]_0/[17]_0$=400) proceeded to 35, 61, and 69% conversion in 10, 30, and 60 min, respectively, and was complete in ca. 80 min.

The GPC profiles showed that the produced polymer exhibited a unimodal, sharp chromatogram, which shifted towards the higher-molecular-weight region as the polymerization proceeded. The Mn of the polymer, as estimated

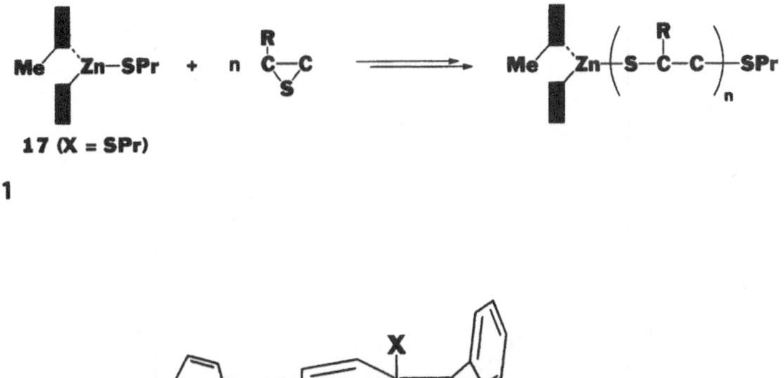

17 (X = SPr)

Scheme 11

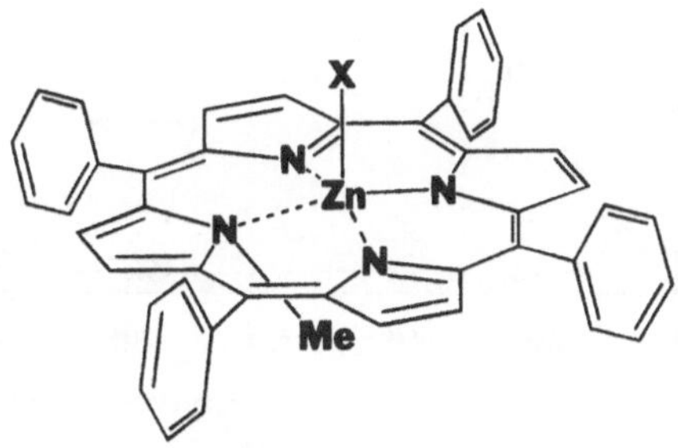

(NMTPP)ZnX

Structure 17

from GPC, increased linearly with the conversion, while the Mw/Mn was almost constant at 1.06 (Fig. 48). Furthermore, the estimated Mn values based on GPC are in good agreement with those (dashed line in Fig. 48) obtained by assuming that every initiator molecule produces one polymer molecule. Accordingly, the Mn of the produced polymer can be controlled over a wide range by changing the mole ratio of the monomer reacted to 17 (X=SPr) ($[PS]_{react}/[17]_0$) retaining the Mw/Mn ratio close to unity. Thus, the polymerization of PS with 17 (X=SPr) proceeds with living character without any chain transfer and termination reactions. This is very different to the polymerization of PS initiated with zinc thiolates $[Zn(SR)_2]$, where in some cases side reactions such as desulfuration of the monomer occur.

Zinc N-phenylporphyrin (17) also brings about the living polymerization of PS, affording a polymer of controlled molecular weight with a narrow MWD. The polymerization with the initial mole ratio $[PS]_0/[17]_0$ of 400 proceeded up to 88% conversion in 50 min at 25 °C, giving a polymer with Mn and Mw/Mn of 24,400 and 1.05, respectively.

The ^{13}C NMR spectrum in CCl_4/C_6D_6 (90:10, v/v) at 60 °C of poly(PS) obtained by the polymerization of PS initiated with 17 (X=SPr) ($[PS]_0/[17]_0=400$,

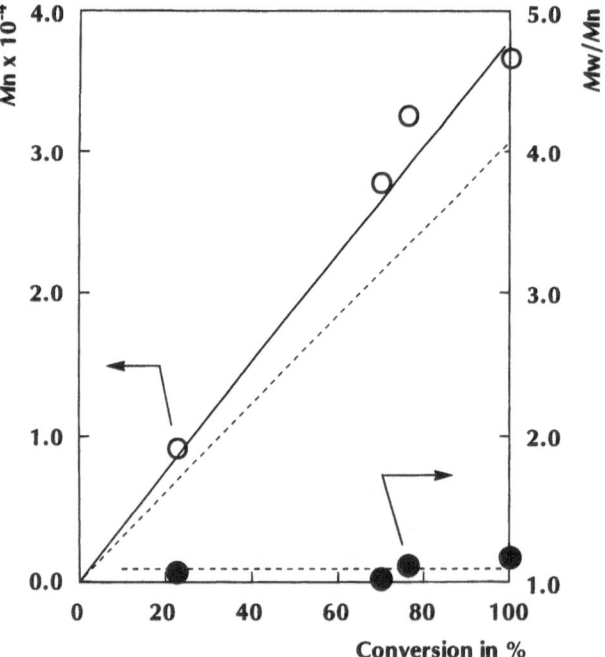

Fig. 48. Polymerization of 1,2-epithiopropane (PS) initiated with (NMTPP)ZnSPr (17, X= SPr) in CH_2Cl_2 at 25 °C ($[PS]_0/[17]_0=400$). Relationship between Mn (O) [Mw/Mn (●)] of the polymer and conversion

69% conversion; $Mn=26,600$ [$Mw/Mn=1.06$]) indicated that the polymer consists of regular head-to-tail sequences. The ratio of the peak areas of the two CH signals indicated the atactic structure of the polymer.

When the polymerization of PS with **17** (X=SPr) was carried out in the presence of thiol as the chain transfer agent, a polymer of uniform molecular weight was obtained with the number of the polymer molecules equal to the sum of those of the molecules of the initiator and thiol (immortal character).

3.2
Visible Light Mediated Living and Immortal Polymerizations of Epoxides Initiated with Zinc Porphyrin [80]

Polymerization of PO with (NMTPP)ZnSPr (**17**, X=SPr) takes place very rapidly under the irradiation with visible light. For example, the reaction did not occur in the dark in 100 min ([PO]$_0$/[**17**]$_0$=40) (Fig. 49), while the polymerization was initiated rapidly upon irradiation and completed in 80 min. It should also be noted that the polymerization, once photoinitiated, did not subside upon turning the light off. As shown by the GPC profiles of the polymerization with the mole ratio [PO]$_0$/[**17**]$_0$ of 430, initiated by irradiation for 40 min, the produced

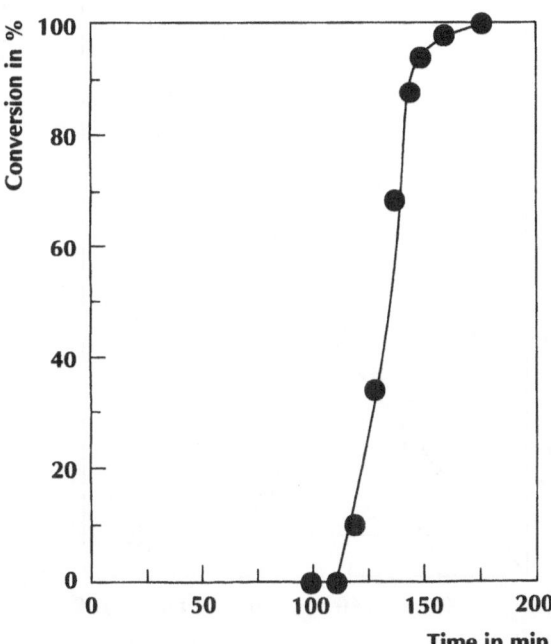

Fig. 49. Polymerization of 1,2-epoxypropane (PO) initiated with (NMTPP)ZnSPr (**17**, X= SPr) ([PO]$_0$/[**17**]$_0$=40) at 26 °C in C$_6$D$_6$ in the dark for the initial 100 min, followed by irradiation (λ>420 nm) for 80 min. Time-conversion curve

polymer exhibited a unimodal, sharp chromatogram, which shifted towards the higher-molecular-weight region as the polymerization proceeded. The Mn of the polymer, estimated from the GPC chromatogram, increased linearly with the monomer conversion, while the Mw/Mn was almost constant at 1.05 (Fig. 50). Furthermore, the Mn values based on GPC are close to those (broken line in Fig. 50) calculated by assuming that every initiator molecule produces one polymer molecule [81]. When 200 equiv of PO were again added to the polymerization system ($[PO]_0/[17]_0$=100; irradiation for the initial 1 h, then in the dark) after the complete consumption of the monomer, the second-stage polymerization ensued in the dark and was complete in 22.5 h. The GPC chromatogram of the polymer clearly shifted from curve to curve retaining the narrow MWD. Thus, the visible light mediated polymerization of PO initiated with (NMTPP)ZnSPr (17, X=SPr) is of a character of *living* polymerization.

The polymerization of PO using 17 (X=SPr) as initiator in benzene ($[PO]_0/[17]_0$=100) took place even in the dark when conducted at 70 °C, affording the polymer with Mn and Mw/Mn of 6100 and 1.07, respectively, after the completion of polymerization (102 min).

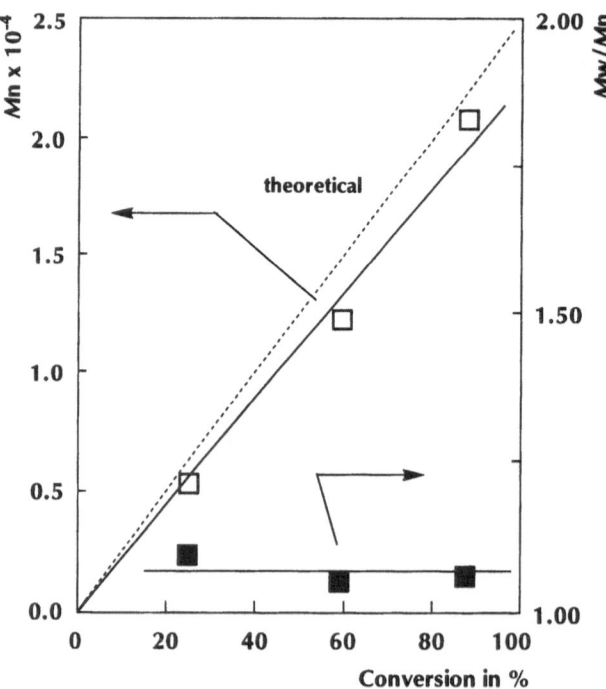

Fig. 50. Polymerization of 1,2-epoxypropane (PO) initiated with (NMTPP)ZnSPr (17, X= SPr) (reaction conditions: see Fig. 49). Relationship between Mn (□) or Mw/Mn (■) and conversion

Poly(PO) formed with **17** as initiator under irradiation showed virtually the same ^{13}C NMR pattern as that for the polymer formed in the dark at 70 °C, where the resonance due to the methyl group was very simple, indicating that the polymer consists of regular head-to-tail linkages. The diad and triad tacticities of the polymer, as determined by ^{13}C NMR [82], indicated the atactic structure: For the polymer formed under irradiation at room temperature, i/s= 0.48:0.52, I/H/S=0.23/0.49/0.28; for the polymer formed in the dark at 70 °C, i/s= 0.49:0.51, I/H/S=0.24/0.50/0.26.

The polymerization of ethylene oxide (epoxyethane, EO) with **17** also proceeded by irradiation with visible light. For example, the polymerization with the mole ratio $[EO]_0/[17]_0$ of 190 in benzene at room temperature, where the monomer conversion after 205 min was very low (<2%, determined by ^1H NMR) in the dark, proceeded to 97% conversion in only 80 min under irradiation. The Mn of the polymer, as estimated from the GPC chromatogram, was 8700, which is in excellent agreement with the expected value of 8100 provided that the numbers of the molecules of the produced polymer and **17** (X=SPr) are equal [81]. The Mw/Mn of the polymer (1.05) was close to unity, indicating the livingness of the visible light induced polymerization of EO initiated with (NMTPP)ZnSPr (**17**).

As described in the above section, the concept of *"immortal"* polymerization in the polymerizations of epoxides and lactones initiated with aluminum porphyrin in the presence of a protic compound has been developed where polymers of uniform molecular weight are formed with the number of the molecules equal to the sum of those of aluminum porphyrin and the protic compound [83]. This is due to the rapid, reversible chain transfer reaction between the molecules of growing polymer and the protic compound, which takes place much faster than the propagation reaction. In the presence of 1-propanethiol (PrSH) as the protic compound, the polymerization of PO ($[PO]_0/[HX]_0/[17$ $(X=SPr)]_0=400/10/1$) proceeded to 93% conversion in 45.5 h at room temperature under irradiation, producing a polymer with Mn and Mw/Mn of 1660 and 1.11, respectively. The number of the polymer molecules relative to that of **17** (X=SPr) (N_p/N_{Zn}), as estimated from Mn and the yield of the polymer [84], is 13.0, being close to the initial mole fraction ($[PrSH]_0+[17]_0)/[17]_0$ of 11. In the ^{13}C NMR spectrum in C_6D_6 of the polymerization mixture, the signals assignable to the propylthio (PrS-) group attached to the polymer terminal [85], the CH_2 group of the terminal unit attached to the PrS- group and the CH group on the other terminal of the polymer chain attached to the OH group were observed in addition to the signals due to the polymer chain. The degree of polymerization of the polymer, as estimated from the intensity ratio of the signals due to CH_2 in the main chain and due to the terminal CH_3 in the PrS- group, was 33.4, which nicely agrees with the mole fraction $[PO]_{reacted}/$ $([PrSH]_0+[17]_0)$ of 33.8. Thus, every polymer molecule carries one propylthio group. The polymerization of PO initiated with **17** (X=SPr) in the presence of methanol (MeOH) also proceeded with *immortal* character under similar conditions, giving at 100% conversion (under irradiation, 18 h) the polymer with

Mn and Mw/Mn of 1680 and 1.08, respectively, and the number of the polymer molecules relative to **17** being 13.8. The ^{13}C NMR spectrum of the polymer showed a signal at δ=58.9 ppm due to the terminal CH_3O group, in addition to one set of relatively weak signals due to the terminal PrS group originating from **17**. The degree of polymerization of the polymer (34.7), as estimated from the intensities of these characteristic signals, was in good agreement with the initial mole fraction $[PO]_0/([MeOH]_0+[17]_0)$ of 36.

3.3
Copolymerization of Epoxide and Episulfide Initiated with Zinc N-Substituted Porphyrin Under Visible Light Irradiation: The First Example of Photoinduced Copolymerizability Enhancement [86]

Copolymerization is a facile method to diversify the structure of polymer materials. However, if the polymerizabilities of comonomers are far from each other, copolymerization is essentially difficult, resulting in the formation of a mixture of the homopolymers and/or the copolymer with block sequences. This is the case for the anionic copolymerization of epoxide and episulfide, where the polymerizability of episulfide is much higher than that of epoxide, and the copolymer consisting mostly of -S-C-C-S- and -O-C-C-O- homo sequences is formed [87]. As mentioned in the previous sections, the zinc complex of N-methylporphyrin brings about polymerization of both epoxide and episulfide.

This section describes that the first example of *photoenhanced copolymerizability* of epoxide and episulfide in a reaction using the zinc complex of N-substituted porphyrin (NMTPP)ZnSPr (**17**, X=SPr) as initiator.

In the copolymerization of EO and PS with (NMTPP)ZnSPr in benzene in the dark at room temperature $([EO]_0/[PS]_0/[(NMTPP)ZnSPr]_0=50/50/1)$, the consumptions of EO and PS took place rather slowly (25 and 80%, respectively, in 30 h), while both were accelerated by elevating the temperature to 70 °C (18 and 85% in 1 h) or irradiation with visible light (57 and 70% in 5.3 h). The consumptions of EO under the three different conditions are plotted against those of PS in Fig. 51, which demonstrates that the relative consumption of EO to PS in the dark at room temperature (●) is far from unity. This tendency is unchanged under accelerating conditions by elevating the temperature to 70 °C in the dark (■). However, of particular interest is that the rates of consumption of the two comonomers come closer to each other under irradiation with visible light (○).

The homopolymerizations of epoxide and episulfide initiated with (NMTPP)ZnSPr (**17**, X=SPr) have been demonstrated to proceed with living character via an (NMTPP)Zn-alcoholate and -thiolate, respectively, as the growing species by the repeated insertions of monomers into the Zn–S or the Zn–O bond of **17** [88,89]. Therefore, the copolymerization of EO and PS with **17** (X= SPr) as initiator is considered to proceed via the following four elementary reactions: the reactions of an (NMTPP)Zn-alcoholate (EO growing end) with EO producing an EO→EO homo sequence (-OCH_2CH_2O-) and with PS producing an EO→PS cross sequence (-$OCH_2CH(CH_3)S$-), and the reactions of an

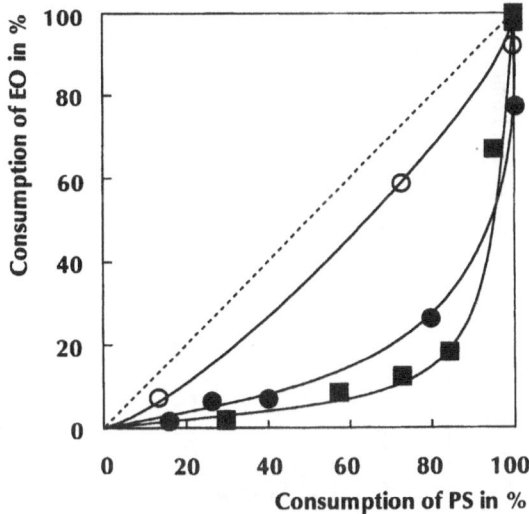

Fig. 51. Copolymerizations of epoxyethane (EO) and 1,2-epithiopropane (PS) initiated with (NMTPP)ZnSPr (**17**, X=SPr) ([EO]$_0$/[PS]$_0$/[**17**]$_0$=50/50/1) in C$_6$D$_6$. Plots of EO consumptions vs. PS consumptions in the dark at rt. (●) in the dark at 70 °C (■), and under irradiation with xenon arc light (λ>420 nm) at rt (□)

(NMTPP)Zn-thiolate (PS growing end) with PS producing a PS→PS homo sequence (-SCH$_2$CH(CH$_3$)S-) and with EO producing a PS→EO cross sequence (-SCH$_2$CH$_2$O-). The contents of these four sequences in the products could be separately evaluated by ^{13}C NMR [90] and the results are summarized in Table 11. In the copolymerizations in the dark (Table 11, runs 1 and 2), the contents of the cross sequences in the products are both only about 10%, while those of the homo sequences are both about 40% irrespective of the copolymerization temperature. The ratios of the contents of the homo and cross sequences [(EO→EO)/(EO→PS), (PS→PS)/(PS→EO)] in the products are thus calculated to be in the range of 3.3 to 4.8. On the other hand, in the product formed under irradiation with visible light, the sum of the contents of the cross sequences is more than 40% with the ratios (EO→EO)/(EO→PS) and (PS→PS)/(PS→EO) being 1.6 and 1.3, respectively (run 3). A similar tendency was observed in the copolymerizations starting from the molar ratio [EO]$_0$/[PS]$_0$/[**1**]$_0$ of 200/200/1 (runs 4 and 5). Thus, by using **17** as initiator for the copolymerization of epoxide and episulfide, the copolymerizability was enhanced by irradiation with visible light.

All the above observations lead to the following conclusions: Irradiation with visible light and elevation of the reaction temperature both result in the accelerated consumptions of comonomers. At 70 °C in the dark, four elementary reactions are almost equally accelerated, taking into account the unchanged sequence distribution and GPC chromatogram of the product compared with the case at room temperature in the dark. On the other hand, irradiation with visible

Table 11. Copolymerization of epoxyethane (EO) and 1,2-epithiopropane (PS) initiated with (NMTPP)ZnSPr (17, X=Spr): contents of homo and cross-sequences in the products[a]

Run	[EO]$_0$/[PS]$_0$/[17]$_0$	Temp.	Light	Content in %[b]				EO→EO	PS→PS
				EO→EO	EO→PS	PS→PS	PS→EO	EO→PS	PS→EO
1[c]	50/50/1	rt[e]	dark	41.3	10.5	37.0	11.2	3.9	3.3
2[c]	50/50/1	70 °C	dark	42.8	8.9	38.4	9.9	4.8	3.9
3[c]	50/50/1	rt[e]	irradiation[f]	31.5	20.3	27.0	21.2	1.6	1.3
4[d]	200/200/1	rt	dark	46.7	3.9	45.1	4.3	12.0	10.5
5[d]	200/200/1	rt	irradiation[f]	41.3	13.1	31.1	14.5	3.2	2.1

a At 100% consumptions of comonomers.
b ^{13}C NMR.
c In C$_6$D$_6$ (5-mm Φ NMR tube).
d In C$_6$H$_6$ (10-ml flask).
e rt (~20 °C).
f Xenon arc light (λ>420 nm).

light is considered to accelerate the cross propagation steps much more than the homo propagation steps, resulting in the contents of the four sequences coming closer to one another, and a change in the GPC chromatogram of the product.

4
Polymerization with a Transition Metal–Porphyrin Complex

4.1
Living and Immortal Ring-Opening Polymerizations of Epoxide Initiated with Manganese Porphyrin [91]

(TPP)MnOAc (**18**) is an excellent initiator of the polymerization of the ring-opening polymerization of PO. An example is illustrated in Fig. 52, where 400 equiv of PO was consumed completely in about 20 h at 30 °C. The ^{13}C NMR spectrum of the polymer shows a simple resonance pattern due to the CH$_3$ group (δ 17.4 ppm), indicating that the polymer consists exclusively of head-to-tail linkages. The ^{13}C NMR spectrum was also informative concerning the stereoregularity of the polyether, where the obtained polyether was almost atactic (i/s= 0.54:0.46, I/H/S=0.28/0.50/0.22). This is in contrast to the case with the aluminum porphyrin as initiator, which produces a polymer rich in isotactic triad sequences under similar conditions.

The gel permeation chromatogram of the polymer obtained with **18** shows a bimodal molecular-weight distribution consisting of a sharp main peak and a

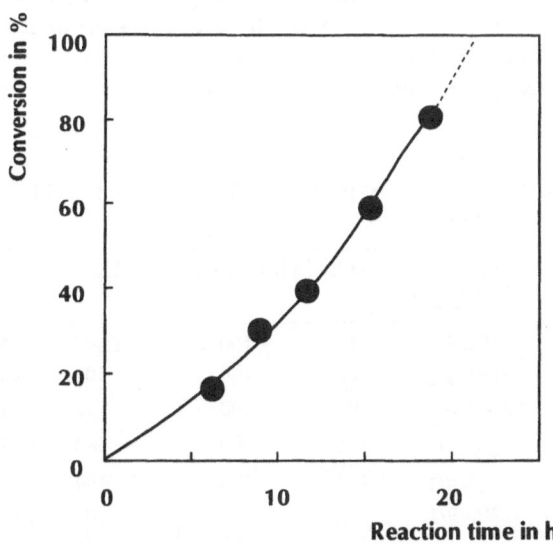

Fig. 52. Polymerization of 1,2-epoxypropane (PO) initiated with (TPP)MnOAc (**18**) ([PO]$_0$/[**18**]$_0$=400, 30 °C, without solvent). Time-conversion relationship

small shoulder on the higher-molecular-weight side. As the polymerization proceeds, the chromatogram of the polymer shifts towards the higher-molecular-weight region without changing the original shape. The molecular weight, corresponding to the top of the main peak (M_{GPC}), and the conversion gave a linear relationship which was almost identical to the calculated line assuming that every molecule of **18** produced one polymer molecule. The polymerization with varying amounts of PO at 100% conversion gave polymers with M_{GPC}s close to the expected values from the initial monomer-to-initiator mole ratios.

When the polymerization of PO (150 equiv, first stage) was carried up to 100% conversion with **18**, followed by adding another 400 equiv of PO (second stage), the GPC profile of the product clearly showed that the chromatogram of the polymer formed at the first stage shifted towards the higher-molecular-weight region, retaining the original shape in the second-stage polymerization. Thus, the polymerization of PO initiated with **18** had a living character, in the sense that the polymer molecules were still alive even after the monomer had been consumed completely.

The IR spectrum of the polymer shows a carbonyl absorption at 1720 cm^{-1} characteristic of an ester group, indicating the incorporation of the axial acetoxy group of **18** into the polymer terminal. Thus, the polymerization is considered to start by insertion of the monomer into the central manganese atom–axial acetoxy group bond, and proceed via a manganese–alkoxide as the growing species.

The polymerization of PO initiated with manganese porphyrin has an immortal nature. Some examples of the polymerization in the presence of methanol (MeOH) as protic compound are listed in Table 12.

The polymerization of PO with **18** in the presence of MeOH ($[PO]_0/[MeOH]_0/[18]_0=400/9/1$) proceeded up to 100% conversion at 30 °C within 48 h (run 1). The Mn and Mw/Mn of the polymer, as estimated from GPC, were 2400 and 1.08, respectively. The number of the polymer molecules relative to that of

Table 12. Polymerization of 1,2-epoxypropane (PO) initiated with (TPP)MnOAc (**18**) in the presence of protic compounds (H-X)[a]

Run	HX	$[PO]_0/$ $[HX]_0/[18]_0$	Time in h	Conversion in %	Mn[b]	Mw/Mn[b]	Np/N_{Mn}[c]
1	CH$_3$OH	400/9/1	48	100	2,400	1.08	9.7
2	CH$_3$OH	400/19/1	60	100	1,100	1.09	21.2
3	CH$_3$OH	400/29/1	96	100	790	1.11	29.4
4	CH$_3$OH	400/39/1	144	100	580	1.07	40.3
5	CH$_3$OH	400/49/1	192	100	480	1.05	48.5
6	CH$_3$COOH	400/9/1	72	100	2,200	1.08	10.5

[a] Without solvent under N$_2$ at 30 °C.
[b] Determined by GPC.
[c] Equal to $0.58 \times [PO]_0 \times [18]_0 \times conversion \times Mn^{-1}$

run 2 (N_p/N_{Mn}), as calculated from the Mn and the yield of the polymer, is 9.7, and is therefore close to the initial mole fraction ($[MeOH]_0+[18]_0)/[18]_0$ of 10. Under similar conditions with varying amounts of MeOH, the Mn of the polymer formed at 100% conversion decreased with the increasing initial mole fraction ($[MeOH]_0+[18]_0)/[18]_0$, while the Mw/Mn remained in the range 1.05–1.11.

4.2
Radical Polymerization of Methyl Methacrylate Initiated with Tin Porphyrin [92]

In contrast to the above polymerizations via anionic and/or coordination anionic mechanisms, radical polymerization initiated with metalloporphyrins remains to be studied. The only example of controlled radical polymerization by metalloporphyrins has been reported by Wayland et al. where the living radical polymerization of acrylic esters initiated with cobalt porphyrins was demonstrated. In this section the radical polymerization of MMA initiated with tin porphyrin is discussed.

In general, the tin–carbon bond in organotin compounds is known to cleave homolytically to give a carbon radical easily. As for the tin complex of porphyrin, it has been reported that when diphenyltin tetraphenylporphyrin (19) in $CHCl_3$ is illuminated with visible light ($\lambda > 420$ nm), phenyl groups are replaced by chlorine [94]. This fact indicates that the tin porphyrin radical generated by the homolytic cleavage of the central tin atom–axial phenyl group bond of 19 is considered to participate in the reaction between 19 and $CHCl_3$.

To a $CHCl_3$ solution of (TPP)SnPh$_2$, prepared from (TPP)SnCl$_2$ and 5 equiv of PhLi in toluene, MMA (200 equiv) was added at room temperature under dry nitrogen in the presence of MeOH (20 eqiuiv). When the mixture was irradiated with visible light ($\lambda > 420$ nm) for 24 h, the viscosity of the mixture increased gradually, indicating that the polymerization of MMA had proceeded. When MeOH was added to the polymerization mixture in order to stop the reaction by precipitating the complex and the products, a polymer was obtained as a MeOH insoluble white powdery product in 10% yield after reprecipitation from $CHCl_3$–MeOH. The Mn of the polymer was estimated as 6500 by GPC ($Mw/Mn=1.5$). The fact that the polymerization of MMA initiated with (TPP)SnPh$_2$ did take place even in the presence of the excess MeOH indicates that the polymerization proceeded via a radical mechanism. On the other hand, the polymerization of MMA did not occur in the dark under otherwise similar conditions, indicating that the visible light irradiation is indispensable for the polymerization.

Acknowledgement. The authors wish to thank Prof. Takuzo Aida of University of Tokyo for his active collaboration and helpful discussions. The authors also thank Drs. Masakatsu Kuroki and Yoshihiko Watanabe, Messrs. Takato Adachi, Masaki Akatsuka, Yasunori Hosokawa, Masanori Isoda, Chikara Kawamura, Masaaki Saika, Daisuke Takeuchi, and Tsuyoshi Watanabe, and Miss Chiho Yamaguchi for their active collaboration.

5
References and Notes

1. Smith KM (ed) (1975) Porphyrins and metalloporphyrins. Elsevier, New York
2. Dolphin D (ed) The porphyrins. Academic Press, New York
3. Ortiz de Montellano PR (ed) (1986) Cytochrome P-450, structure, mechanism and biochemistry. Plenum Press, New York
4. Mansuy D (1987) Pure Appl Chem 59:759
5. Osa T (ed) (1982) Porufirin no Kagaku. Kyoritsushuppan, Tokyo
6. Rix CJ (1982) J Chem Educ 59:389
7. Poulos TL (1988) Heme Proteins. In: Eichhorn GL, Marzill LG (eds) Adv Inorg Biochem 7:1–36
8. White RE, Coon MJ (1980) Annu Rev Biochem 49:315
9. (a) Deisenhofer J, Epp O, Miki K, Huber R, Michael H (1985) Nature 318:618; (b) Allen JP, Feher, G, Yrates TO, Komiya H, Rees DC (1987) Proc Natl Acad USA 84:5730
10. (a) Woodward RB (1973) Pure Appl Chem 33:145
11. (a) Collman JP, Gagne RR, Halbert TR, Marchon J-C (1973) J Am Chem Soc 95:7868; (b) Alworg A, Baldwin JE, Huff J (1975) J Am Chem Soc 97:27; (c) Ogoshi H, Sugimoto H, Yoshida Z (1976) Tetrahedron Lett 4477; (d) Baldwin JE, Klose T, Peters M (1976) J Chem Soc Chem. Commun 881; (e) Collman JP, Brauman JI, Doxsee KM, Halbert TR, Bunnenberg E, Linder RE, Lamar GN, Gaudio JD, Lang G, Spartalian KJ (1980) Am Chem Soc 102:4182; (f) Chang, CK (1977) J Am Chem Soc 99:2819; (g) Chang CK, Traylor TG (1973) Proc Natl Acad Sci USA 50:951
12. (a) Collman JP, Suslick KS (1978) Pure Appl Chem 50:951; (b) Traylor TG Acc Chem Rev 1981, 14, 102; (c) Yoshida Z (1984) Heterocycles 21:331; (d) Tsuchida E, Nishide H (1986) Top Curr Chem 132:63; (e) Tabushi I (1988) Coord Chem Rev 86:1
13. Aida T, Mizuta R, Yoshida Y, Inoue S (1981) Makromol Chem 182:1073
14. (a) Yasuda T, Aida T, Inoue S (1982) Makromol Chem Rapid Commun 3:585; (b) Yasuda T, Aida T, Inoue S (1984) Macromolecules 17:2217; (c) Endo M, Aida T, Inoue S (1987) Macromolecules 20:2982; (d) Shimasaki K, Aida T, Inoue S (1987) Macromolecules 20:3076; (e) Trofimoff LR, Aida T, Inoue S (1987) Chem Lett 991; (f) Sugimoto H, Aida T, Inoue S (1990) Macromolecules 23:2869
15. Kuroki M, Aida T, Inoue S (1987) J Am Chem Soc 109:4737
16. Hosokawa Y, Kuroki M, Aida T, Inoue S (1991) Macromolecules 24:824
17. Kuroki M, Watanabe T, Aida T, Inoue S (1991) J Am Chem Soc 113:5903
18. Sugimoto H, Kuroki M, Watanabe T, Kawamura C, Aida T, Inoue S (1993) Macromolecules 26:3403
19. Sugimoto H, Aida T, Inoue S (1993) Macromolecules 26:4751
20. For brief reviews: (a) Sugimoto H, Aida T, Inoue S (1995) Bull Chem Soc Jpn 68:1239; (b) Sugimoto H, Inoue S (1998) Polymer News, in press
21. The assignment was made according to: Maruoka K, Araki Y, Yamamoto HJ (1988) J Am Chem Soc 110:2650 (Supplementary material)
22. ^{1}H NMR for 2 (R=^{t}Bu) (C$_6$D$_6$): δ 9.36 (pyrrole-β-H), 8.46 (Ph-o-H), 7.76 (Ph-m, p-H), 2.7–2.3 (main chain CH$_2$), 1.9–1.4 (main chain CH$_3$ and C(CH$_3$)$_3$), 0.96 (terminal CH$_3$; e in Fig. 9), 0.72 and 0.70 (=CCH$_3$; b), 0.51 (=C(CH$_3$)CH$_2$CCH$_3$; d), 0.43–0.40 (=CCH$_2$; c), 0.0 – -0.2 (=C(CH$_3$)CH$_2$CCH$_3$; d), -0.28 and -0.33 (=COC(CH$_3$)$_3$; a), +1.00 and -1.09 (=CCH$_3$; b), -1.14 (=CCH$_2$; c)
23. A small shoulder was observed on the higher-molecular-weight side of the main peak in GPC
24. Sugimoto H, Aida T, Inoue S (1994) Macromolecules 27:3672
25. Sugimoto H, Aida T, Inoue S (1993) Macromolecules 26:4751
26. From the relative intensity of the OCH$_3$ signals of MMA (δ 3.75 ppm) to the polymer (δ 3.65)

27. The green color is typically observed for the axial ligand-exchange reaction of (TPP)AlCl with KF (unpublished results)
28. Sugimoto H, Kuroki M, Watanabe T, Kawamura C, Aida T, Inoue S (1998) Macromolecules in press
29. (a) Kuroki M, Watanabe T, Aida T, Inoue S (1991) J Am Chem Soc 113:5903; (b) Adachi T, Sugimoto H, Aida T, Inoue S (1992) Macromolecules 25:2280; (c) Adachi T, Sugimoto H, Aida T, Inoue S (1993) Macromolecules 26:1238
30. Guilard R, Zrineh A, Tabard A, Endo A, Hau BC, Lecomte C, Souhassou M, Habbou A, Ferhat M, Kadish KM (1990) Inorg Chem 29:4476. ^{1}H NMR in C_6D_6 [C_6H_6 (δ 7.40) as internal standard]: δ 9.34 (pyrrole-β-H, 8H, s), 8.27 (Ph-o-H, 8H, d), 7.68 (Ph-m, p-H, 12H, m), 6.08 (Al-Ph-p-H, 1H, t), 5.83 (Al-Ph-m-H, 2H, t), and 2.73 (Al-Ph-o-H, 2H, t)
31. Adachi T, Sugimoto H, Aida T, Inoue S (1993) Macromolecules 26:1238
32. 1c also initiates the living polymerization of *tert*-butyl acrylate even in the dark: Hosokawa Y, Kuroki M, Aida T, Inoue S (1992) Macromolecules 25:824
33. Adachi T, Sugimoto H, Aida T, Inoue S (1993) Macromolecules 26:1238
34. The assignments were made based on the ^{13}C NMR spectra for $(CH_3CH_2CH_2)_2S$, δ 34.3 (SCH_2), 23.3 (CH_2Me), and 13.6 (CH_3), and for $CH_3CH_2CH_2SH$, δ 27.4 (SCH_2), 26.6 (CH_2Me), and 12.9 (CH_3), in C_6D_6
35. Ute K, Nishimura T, Hatada K (1989) Polymer J 21:1027
36. The assignments were made by reference to the ^{1}H NMR spectrum in CCl_4 of methyl 2-methyl-3-propylthiopropionate [$CH_3CH(CO_2CH_3)CH2SCH2CH2CH_3$]: δ 0.95 (CH_2CH3), 1.33 (CH_2Me), 2.36 (SCH_2Et), and 2.51 and 2.72 (CH_2SPr): Georges G, Rouvier E, Musso J, Cambon A, Fellous R (1972) Bull Soc Chim Fr 4622
37. The selection of the wavelength for UV detection was based on the λ_{max} of thioanisole in THF
38. Even when PMMA prepared with 1a was mixed with benzenethiol prior to GPC analysis, no UV response was observed for the polymer fraction, where benzenethiol eluted very late and the peak was perfectly separated from that of the polymer
39. Adachi T, Sugimoto H, Aida,T, Inoue S (1992) Macromolecules 25:2280
40. Adachi T, Sugimoto H, Aida T, Inoue S (1992) Macromolecules 25:2280
41. Sugimoto H, Saika M, Hosokawa,Y, Aida,T, Inoue S (1996) Macromolecules 29:3359
42. Feit B-A, Heller E, Zilkha A (1966) J Polym Sci Part A-1 4:1151
43. Beaman RC (1948) J Am Chem Soc 70:3115
44. Rempp P, Blumstein A (1961) Bull Soc Chim Fr 1018
45. Overberger CG, Yuki H, Urakawa N (1960) J Polym Sci 45:127
46. Feit B-A, Mirelman D, Zilkha A (1964) J Polym Sci Part A-1 2:4743
47. Overberger CG, Pearce EM, Mayes N (1958) J Polym Sci 31:217
48. Zilkha A, Feit B-A, Frankel M (1961) J Polym Sci 49:231
49. Feit B-A, Direlman D, Zilkha A (1965) J Appl Polym Sci 9:2459
50. Webster OW, Hertler WR, Sogah DY, Farnham WB, RajanBabu TV (1983) J Am Chem Soc 105:5706
51. (a) Sugimoto H, Aida T, Inoue S (1995) Bull Chem Soc Jpn 68:1239; (b) Kuroki M, Watanabe T, Aida T, Inoue S (1991) J Am Chem Soc 113:5903; (c) Sugimoto H, Kuroki M, Watanabe T, Kawamura C, Aida T, Inoue S (1993) Macromolecules 26:3403; (d) Adachi T, Sugimoto H, Aida T, Inoue S (1992) Macromolecules 25:2280; (e) Adachi T, Sugimoto H, Aida T, Inoue S (1993) Macromolecules 26:1238
52. (a) Takeda N, Inoue S (1977) Bull Chem Soc Jpn 50:984; (b) Aida T, Inoue S (1983) J Am Chem Soc 105:1304; (c) Hirai Y, Aida T, Inoue S (1989) J Am Chem Soc 111:3062
53. (a) Bovey FA, Tiers GVD (1960) J Polym Sci 44:173; (b) Yuki H, Hatada K, Nimori T, Kikuchi Y (1970) Polym J 1:36
54. The effect of visible light on the propagating step was also observed for the polymerization of MMA [12]; however, the acceleration effect for the polymerization of MAN was much more remarkable
55. Aida T, Mizuta R, Yoshida Y, Inoue S (1981) Makromol Chem 182:1073

56. (a) Yasuda T, Aida T, Inoue S (1982) Makromol Chem Rapid Commun 3:585; (b) Yasuda T, Aida T, Inoue S (1984) Macromolecules 17:2217; (c) Endo M, Aida T, Inoue S (1987) Macromolecules 20:2982; (d) Shimasaki K, Aida T, Inoue S (1987) Macromolecules 20:3076; (e) Trofimoff LR, Aida T, Inoue S (1987) Chem Lett 991; (f) Sugimoto H, Aida T, Inoue S (1990) Macromolecules 23:2869
57. Kuroki M, Aida T, Inoue S (1987) J Am Chem Soc 109:4737
58. Hosokawa Y, Kuroki M, Aida T, Inoue S (1991) Macromolecules 24:824
59. (a) Kuroki M, Watanabe T, Aida T, Inoue S (1991) J Am Chem Soc 113:5903; (b) Sugimoto H, Kuroki M, Watanabe T, Kawamura C (1993) Macromolecules 26:3403
60. Sugimoto H, Kuroki M, Kawamura C, Aida T, Inoue S (1994) Macromolecules 27:2013
61. N_p/N_{TPP}={molecular weight of PO (58)∞[PO]$_0$/[2]$_0$$\infty$(conversion/100)}/Mn
62. (a) Vincens V, Le Borgne A, Spassky N (1989) Makromol Chem Rapid Commun 10:623; (b) Vincens V, Le Borgne A, Spassky N (1992) In: Vandenberg EJ, Salamone JC (eds) Catalysis in polymer synthesis. ACS Symp Ser No. 496, chap 16, p 205; (c) Le Borgne A, Vincens V, Jouglard M, Spassky N (1993) Makromol Chem Macromol Symp 73:37
63. Akatsuka M, Aida T, Inoue S (1994) Macromolecules 27:2820
64. For immortal polymerizations: (a) Asano S, Aida T, Inoue S (1985) J Chem Soc Chem Commun 1148; (b) Aida T, Maekawa Y, Asano S, Inoue S (1988) Macromolecules 21:1195; (c) Endo M, Aida T, Inoue S (1987) Macromolecules 20:2982
65. (a) Thanabal V, de Ropp JS, La Mar GN (1987) J Am Chem Soc 109:265; (b) Arai T, Inoue S (1990) Tetrahedron 46:749
66. Takeuchi D, Watanabe Y, Aida T, Inoue S (1995) Maclomolecules 28:651
67. Recent reviews: (a) Aida T (1994) Prog Polym Sci 19:469; (b) Inoue S, Aida T (1998) Controlled polymer synthesis with metalloporphyrins. In: Vogl O, Hatada K (eds) Molecular design of polymeric materials. Dekker, New York, in press
68. (a) Kuroki M, Watanabe T, Aida T, Inoue S (1991) J Am Chem Soc 113:5903; (b) Sugimoto H, Kuroki M, Watanabe T, Kawamura C, Aida T, Inoue S (1993) Macromolecules 26:3403; (c) Sugimoto H, Aida T, Inoue S (1994) Macromolecules 27:3672; (d) Sugimoto H, Kawamura C, Kuroki M, Aida T, Inoue S (1994) Macromolecules 27:2013; (e) Akatsuka M, Aida T, Inoue S (1994) Macromolecules 27:2820
69. Inoue, S, Aida T (1994) Chemtech 24:28
70. A part of the present work was presented at the 43rd Annual Meeting of the Society of Polymer Science, Japan (May 1994, Nagoya): Polymer Prepr. Jpn. 1994, 43, 171. Concurrently, Amass et al. reported the polymerization of oxetane with (TPP)AlCl (1) alone at 55 °C: Amass AJ, Perry MC, Riat DS, Tighe BJ, Colclough E, Steward M (1994) J Eur Polym J 30(5):641
71. Mixing of oxetane (15 mmol) with 4 (0.3 mmol) in CH_2Cl_2 (3 ml) in the absence of 1 under similar conditions resulted in a polymerization of oxetane, where the monomer conversion was stopped at 61.4% (2 h), affording a broad MWD polymer (Mn~10^5, Mw/Mn=1.93) together with some cyclic oligomers
72. Narrow MWD polyoxetanes prepared with the 2–4 system, having defined Mns by ^{1}H NMR end group [OCH(CH$_3$)$_2$] analysis, were used as standards for GPC calibration for evaluating Mns and Mw/Mn ratios of other polyoxetane samples
73. Isoda M, Sugimoto H, Aida T, Inoue S (1997) Macromolecules 30:57
74. (a) Yasuda T, Aida T, Inoue S (1982) Makromol Chem Rapid Commun 3:585; (b) Yasuda T, Aida T, Inoue S (1986) Bull Chem Soc Jpn 59:3931; (c) Sugimoto H, Aida T, Inoue S (1990) Macromolecules 23:2869
75. (a) Endo M, Aida T, Inoue S (1987) Macromolecules 20:2982; (b) Shimasaki K, Aida T, Inoue S (1987) Macromolecules 20:3076
76. For transesterification of polylactone, see: (a) Nobutoki K, Sumitomo H (1967) Bull Chem Soc Jpn 40:1741; (b) Perret R, Skoulious A (1972) Makromol Chem 152:291; (c) Kricheldorf HR, Mang T, Jont JM (1984) Macromolecules 17:2173; (d) Kricheldorf HR, Berl M, Scharnagl N (1988) Macromolecules 21:286; (e) Kricheldorf HR, Kreiser I

(1987) J Macromol Sci Chem A24(11):1345; (f) Kricheldorf HR, Kreiser I, Scharnagl N (1990) Makromol Chem Macromol Symp 32:285

77. Poly(δ-VL) was synthesized by the polymerization of δ-VL with the (TPP)AlOMe–**2a** system (Mn=24,000 [Mw/Mn=1.57]), purified by precipitation from $CHCl_3$–hexane, and subjected to ^{13}C NMR studies

78. The signals were assigned by reference to the signals for the model compound such as (TPP)Al-O-CH(CH$_3$)-CH$_2$-C(=O)-OEt (**1**, X=O-CH(CH$_3$)-CH$_2$-C(=O)-OEt) prepared by the reaction of (TPP)AlMe (**1**, X=Me) and HO-CH(CH$_3$)-CH$_2$-C(=O)-OEt

79. Aida T, Kawaguchi K, Inoue S (1990) Macromolecules 23:3887

80. Watanebe Y, Aida T, Inoue S (1990) Macromolecules 23:2612

81. $M_{calc}=\{[monomer]_0/[17]_0\}\times MW(monomer)\times conversion/100)$

82. The distributions of isotactic (I) and syndiotactic (s) diad together with isotactic (I), heterotactic (H) and syndiotactic (S) triad sequences were determined from the resonances due to methylene and methine carbons: Oguni N, Lee K, Tani H (1972) Macromolecules 5:819

83. (a) Asano S, Aida T, Inoue S (1985) J Chem Soc Chem Commun 1148; (b) Aida T, Maekawa Y, Asano, S, Inoue S (1988) Macromolecules 21:1195; (c) Inoue, S, Aida T (1986) Makromol Chem Macromol Symp 6:217

84. $N_p/N_{Zn}=([monomer]_0/[17]_0)\times conversion/100\times MW(monomer)/Mn$

85. The assignments were made based on the ^{13}C NMR spectra for (CH$_3$CH$_2$CH$_2$)$_2$S; δ 34.3 (SCH$_2$), 23.3 (CH$_2$Me), 13.6 (CH$_3$), and CH$_3$CH$_2$CH$_2$SH; δ 27.4 (SCH$_2$), 26.6 (CH$_2$Me), 12.9 (CH$_3$) in C$_6$D$_6$

86. Watanebe Y, Aida T, Inoue S (1991) Macromolecules 24:3970

87. (a) Sigwalt P, Spassky N (1984) In: Ivin KJ, Saegusa T (eds) Ring-opening polymerization, vol 2. Elsevier, London, p 685; (b) Kuznetsov, Yu P, Glumova TD, Polotskaya GA, Belonovskaya GP (1981) Vysokomol. Soedin Ser A 23:2217

88. Watanabe Y, Aida T, Inoue S (1990) Macromolecules 23:2612

89. Aida T, Kawaguchi K, Inoue S (1990) Macromolecules 23:3887

90. ^{13}C NMR in C$_6$D$_6$ [C$_6$D$_6$ (δ128.0) as internal standard]: For EO'EO (-OCH$_2$CH$_2$O-), 70.9 (CH$_2$); EO'PS (-OCH$_2$CH(CH$_3$)S-), δ76.8 (CH$_2$), 40.1 (CH), and 18.8 (CH$_3$); PS'PS (-SCH$_2$CH(CH$_3$)S-), 41.6 (CH), 38.8 (CH$_2$), and 20.9 (CH$_3$); and PS'EO (-SCH$_2$CH$_2$O-), 30.7 (SCH$_2$) and 71.8 (CH$_2$O). The assignment of the signals was made based on the spectra of the following compounds: Poly-EO, 71.0 (CH$_2$); poly-PS, δ 41.5 (CH), 38.8 (CH$_2$), and 20.9 (CH$_3$); EtOCH$_2$CH(CH$_3$)SEt, 76.2 (CH$_2$), 39.5 (CH), and 18.7 (CH$_3$); and EtSCH$_2$CH$_2$OEt, 31.5 (SCH$_2$) and 71.1 (CH$_2$O)

91. Kuroki M, Aida T, Inoue S (1988) Makromol Chem 189:1305

92. Yamaguchi C, Sugimoto H, Inoue S unpublished results

93. Wayland BB, Poszmik G, Mukerjee SL, Fryd MJ (1994) J Am Chem Soc 116:7493

94. Dawson DY, Sangalang JC, Arnold J (1996) J Am Chem Soc 118:6082

Editor: Prof. A. Abe
Received: July 1998

Interpolymer Complexation and Miscibility Enhancement by Hydrogen Bonding

Ming Jiang*, Mei Li, Maoliang Xiang, Hui Zhou

Institute of Macromolecular Science and Laboratory of Molecular Engineering of Polymers, Fudan University, Shanghai 200433, China
* e-mail: mjiang@fudan.ac.cn

The main development in macromolecular complexation in aqueous media due to hydrogen bonding over the last decade has been a better understanding of the process of complex formation. Meanwhile an ever growing interest has occurred in introducing hydrogen bonding into the polymers which originally lack donor or acceptor groups. For such polymer solutions it has been demonstrated that the transition from separated coils to complex aggregates takes place when the content of introduced interaction sites reaches a certain level. For the blends with controllable hydrogen bonding in the solid state, the relevant experiments have shown that immiscibility-miscibility-complexation transitions occur upon progressive increase in the density of hydrogen bonding.

Keywords: Interpolymer complexation, Hydrogen bonding, Miscibility, Miscibility-complexation transition, Fluorospectroscopy, Complex aggregate

Advances in Polymer Science, Vol.146
© Springer-Verlag Berlin Heidelberg 1999

List of Abbreviations

AM   acrylamide
BVPy   poly(butyl methacrylate-co-4-vinyl pyridine)
CPS   partially carboxylated polystyrene
DAAm   *N,N*-dimethylacrylamide
DMAc   *N,N*-dimethylacetamide
DMF   *N,N*-dimethylformamide
DMSO   dimethyl sulfone

HEMA	hydroxyethyl methacrylate
HES	p-(1-hydroxy ethyl) styrene
HFMS	p-(1,1,1,3,3,3-hexafluoro-2-hydroxypropyl)-α-methyl styrene
HFS	p-(1,1,1,3,3,3-hexafluoro-2-hydroxypropyl) styrene
HPS	p-(2-hydroxyisopropyl) styrene
HPAM	partially hydrolyzed PAM
i-PMMA	isotactic PMMA
IPN	interpenetrating polymer network
LCST	low critical solution temperature
LLS	laser light scattering
MPαMS	PαMS modified with hydroxyl-containing monomer unit
NMI	N-maleimide
NRET	non-radiative energy transfer
P2VPy	poly(2-vinyl pyridine)
P4VPy	poly(4-vinyl pyridine)
PAA	poly(acrylic acid)
PAM	poly(acrylamide)
PAMA	poly(alkyl methacrylate)
PαMS	poly(α-methyl styrene)
PBA	poly(butyl acrylate)
PBMA	poly(butyl methacrylate)
PC	polycarbonate
PCL	polycaprolactone
PDMA	poly(N,N-dimethyl acrylamide)
PEA	poly(ethyl acrylate)
PEMA	poly(ethyl methacrylate)
PEO	poly(ethylene oxide)
PEOX	poly(ethyl oxazole)
PES	poly(ethyl succinate)
PHMP	poly[(1-hydroxy-2,6-phenylene)methylene]
PI-b-PMMA	poly(isoprene-b-methyl methacrylate)
PI(OH)	copolymer of isoprene and HFMS
PMAA	poly(methylacrylic acid)
PMBI	poly(monobenzyl itaconate)
PMEI	poly(monoethyl itaconate)
PMMI	poly(monomethyl itaconate)
PMPMA	poly(N-methyl-4-piperidinyl methacrylate)
PMVAc	poly[N-methyl-N-vinylacetamide]
PPF	p-bromophenol-formaldehyde copolymer
PPO	poly(2,6-dimethylphenylene oxide)
PS	polystyrene
PS(OH)	copolymer of styrene and HFMS
PS(s-OH)	poly{styrene-co-[p-(1-hydroxyethyl) styrene]}
PS(t-OH)	poly{styrene-co-[p-(2-hydroxypropan-2-yl] styrene}

PSe-OH   polystyrene with end group of (2,2,2-trifluoro-1-hydroxy-1-
            trifluoromethyl ethyl)
PSF   polysulfone
PSVBDEP   poly(styrene-co-4-vinyl benzene phosphonic acid diethyl ester)
PVAc   poly(vinyl acetate)
PVME   poly(vinyl methyl ether)
PVPh   poly(vinyl phenol)
PVPo   poly(vinyl pyrrolidone)
s-PMMA   syndiotactic PMMA
SAA   poly(styrene-co-allyl alcohol)
SBS   poly(styrene-b-butadiene-b-styrene)
SEBS   poly[styrene-b-(ethylene-co-butylene)-b-styrene]
SIS   poly(styrene-b-isoprene-b-styrene)
SPS   partially sulfonated polystyrene
STHFS   copolymer of styrene and HFS
STVPh   poly(styrene-co-vinyl phenol)
STVPy   poly(styrene-co-4-vinyl pyridine)
ST2VPy   poly(styrene-co-2-vinyl pyridine)
TEM   transmission electron microscopy
VBA   4-vinyl benzoic acid
VPDMS   4-vinyl phenyldimethylyl silanol
VPh   vinyl phenol

1
Introduction

In the literature, the term "macromolecular complexes" is used to refer to two different objects: macromolecular metal complexes and intermacromolecular or polymer–polymer complexes. The first object contains macromolecules and metal ions that provide organic polymers with inorganic functions. The formation and structure of the complexes of this type and particularly their dynamic interactions and electronic processes are summarized in Tsuchida's book published in 1991 [1]. The second object is actually an intermolecular associate of two different polymers bound together by secondary binding forces. It is usually divided into four classes, known as polyelectrolyte [2–5], hydrogen-bonding [2–4], stereo [6] and charge-transfer [7] complexes, and the first two have received most of the attention of researchers. In this review, we limit our discussion to the complexation of synthetic polymers due to hydrogen bonding, and do not cover as many of the topics that the two review articles published in the early 1980s did [2,3].

The development made on polymer complexation in aqueous media since these reviews has been mainly directed at a better understanding of its process by the use of more sophisticated techniques, rather than adding more polymer systems to the investigation. Therefore, instead of summarizing the complexation of a variety of polymers comprehensively, we focus on some typical topics important for exploring the process and mechanism of polymer complexation. Section 2 deals with these subjects in relation to water-soluble polymers. While the early studies chiefly directed interest to such polymers, the trendy subject over the past decade has been polymer complexation in non-aqueous media. Section 3 discusses the results of these investigations.

The work reviewed in Sects. 2 and 3 mostly concerns pairs of polymer components each having either a proton-donor or proton-acceptor group as its inherent part. However, over the past decade, an ever growing interest has occurred in introducing hydrogen bonding into the polymers which originally lack suitable donor or acceptor groups, and much work has been done aimed at enhancing miscibility of otherwise immiscible blends or, in other words, realizing an immiscibility-miscibility transition. This topic is briefly reviewed in Sect. 4. Although work of this kind is not directly concerned with complexation, it has raised the interesting question as to whether complexation can be caused simply by strengthening the hydrogen bonding in a given system. Thus, over the past decade, work has been done to incorporate more hydrogen bonds than just necessary for achieving miscibility, and the results have shown that, in so doing, ordinary miscible blends can be converted into complex blends. In addition, the related study of mixing the polymer components with controllable hydrogen bonding in solution has provided clear evidence for the complexation and its dependence on the structure parameters. The complexation of polymer systems with introduced hydrogen bonding in solution and also in the bulk state is reviewed in Sects. 5 and 6, respectively.

2
Interpolymer Complexation in Aqueous Media

Interpolymer complexation between water-soluble polymers by hydrogen bonding was a frontier subject in the 1970s. Poly(carboxylic acids), mainly poly(acrylic acid) (PAA) and poly(methacrylic acid) (PMAA), served as the most common proton-donating components. As for the proton-accepting polymers, poly(ethylene oxide) (PEO or PEG) and poly(N-vinyl-2-pyrrolidone) (PVPo) were often used. The important results on the formation of complex aggregates and its dependence on the structur ˇ prameters have been reviewed [2,3,8]. In this section we select a few representative topics to look at recent advances in interpolymer complexes in aqueous media, with the emphasis on fluorescence probe studies.

2.1
Thermodynamics of Interpolymer Complexes

The thermodynamics of the complexation of a typical polymer pair consisting of proton-donating polyacid and proton-accepting poly(ethylene oxide) (PEO) has been studied since the late 1970s [9,10]. For complexes formed by cooperative hydrogen bonding between a pair of polyacid and polybase with stoichiometric composition, the stability constant K for the equilibrium between hydrogen bonded and free sites is given by

$$K=\theta/C_0 (1-\theta)^2 \tag{1}$$

where θ is the degree of conversion defined as the fraction of bonded carboxylic groups and C_0 is the initial concentration of the polyacid in repeating units. If the apparent dissociation constant of the polyacid is assumed not to vary with complexation, θ is given by

$$\theta=1-([H^+]/[H^+]_0)^2 \tag{2}$$

where $[H^+]$ and $[H^+]_0$ are the hydrogen ion concentrations in the presence and absence of complementary polymers, respectively. Therefore, by measuring the pH of the complex solution, it is possible to determine both θ and K. Furthermore, by determining the temperature dependence of K, the basic thermodynamic parameters can be calculated from the well-known equations:

$$\Delta F^0=-RT \ln K \tag{3}$$

$$d(\ln K)/d(1/T)=-\Delta H^0/R \tag{4}$$

$$\Delta S^0=-(\Delta F^0-\Delta H^0)/T \tag{5}$$

In their studies on the complexes PMAA/PEO and PAA/PEO, Tsuchida et al. [10] found that ΔH^0 and ΔS^0 strongly depended on the molecular weight of PEO; see also a recent review by Tsuchida and Takeoka [4]. The thermodynamics for the complexation of PAA with PVPo was recently studied by pH measurements at different temperatures [11], and it was shown that ΔH^0 and ΔS^0 were positive at lower temperature and decreased continuously with increasing temperature. The large positive ΔH^0 values (70–80 Kcal/mol) at lower temperatures were interpreted as due to hydrophobic interactions and conformational changes during complexation, and the positive ΔS^0 values were considered as reflecting the release of water during complexation. It was reported [12] that this method of determining the thermodynamic parameters could also be successfully applied to the complexation of PVPo with copolymers of methacrylic acid–methacrylamide and acrylic acid–acrylamide if the concentration C_0 was simply identified to the concentration of active units in the copolymers.

The kinetics and equlibria of the complexation between PAA and PEO or PVPo were studied by Morawetz's group [13–15], and it was shown that the complex formation consisted of an initial diffusion-controlled hydrogen-bonding process with a small activation energy and an extensive conformational transition of the two polymer chains which induces additional hydrogen bonding, thus stabilizing the complex.

2.2
Effect of Molar Mass of Polymers on Complexation

It is generally accepted that, because of the cooperative nature of the interaction, no complexes are formed when the molar masses of the interacting polymers are less than a "critical value". Tsuchida and Abe [2] summarized some of the experimental critical values; for example, the critical degrees of polymerization of PEO for complexation with PAA and PMAA are approximately 200 and 40, respectively [16,17]. Iliopoulos and Audebert [18] proposed a simple method for estimating the minimal chain length (m.c.l.) for PAA complexation. By partially hydrolyzing PAA, they prepared a series of modified PAA having different average AA unit sequences and measured the pH for their complexes with PEO in solution, to determine θ and m.c.l. For example, at a PAA concentration C=0.02 unit mol/L and a unit molar ratio [PEO]/[PAA]=1, the m.c.l. was about 7. It appears that this value is much less than the m.c.l. of PEO.

Using dansyl-labeled PAA (Dan-PAA), Chen and Morawetz [14] and Bednár et al. [19] investigated the effect of molecular weight on the complexation of PAA with PEO or PVPo by fluorospectroscopy. This method takes advantage of the sensitivity of the emission characteristic of the dansyl group to its microenviroment. In fact, the fluorescence is about one order of magnitude more intense in organic solvents than in water, so that complexation of PAA with proton-accepting polymers may be monitored from the intensity increase caused by the displacement of water molecules from the neighborhood of the label. When Dan-PAA of 1.4×10^5 molecular weight was mixed with PEO of molecular weights ranging from 3400 to 24,000 or PVPo of molecular weights ranging from 10,000 to 410,000 at different pH and NaCl concentrations, it was observed that the higher the molecular weights of the proton-accepting polymers, the stronger the interaction became; this is in good agreement with the observations of many others [11]. It was also observed that although PAA/PVPo showed a 1:1 stoichiometry for higher molecular weight PVPo, PAA/PEO did not show a definite stoichiometry. In fact, PVPo was found to form more stable complexes with PAA than PEO did, consistent with the literature [10,20–22].

When complexation of PAA with much higher molecular weight PEO was studied by measuring fluorescent emission of Dan-PAA, some unexpected results were obtained [19]. In the solutions containing Dan-PAA and its counterpart PEO-(16,000) or PEO-(24,000), the intensity increased steadily as the PEO/PAA base mole ratio increased, reflecting the formation of a complex. For both PEOs, the emission exhibited a blue shift, but the addition of much higher molecular weight PEO (3.1×10^5 and 24.3×10^5) induced a red shift with no apparent intensity increase. This can be taken as indicating an increased exposure of the dansyl label to water. Based on these and some additional experimental data, Bednár et al. [19] proposed the following. In the complexes of Dan-PAA with shorter chain PEO, the dansyl label has efficient access to the PEO and creates a more hydrophobic environment, while in the complexes with larger chain PEO, the Dan-PAA chain can contact the PEO only at widely separated points along its contour and the PAA chains between these points are stretched out. Thus, some of the dansyl groups are removed from the aqueous environment or even more directly contact water molecules than unassociated Dan-PAA chains. It is interesting to note that this work indicates that not only the ability of complexation but also the structure of the complex aggregates can be significantly affected by the molecular weights of the component polymers.

Oyama et al. [23] utilized fluorescence spectroscopy to study how the molar masses of both PAA and PEO affect complexation. Actually, they used the excimer formation between the pyrene groups attached to the chain ends of PEO as a molecular probe. The molecular weights of PAA used were 1850, 4600 and 890,000, and those of PEO were 4800 and 9200. The latter are much lower than the PEO molecular weight in the work of Bednár et al. [19]. In order to monitor both *intra*molecular and *inter*molecular excimer formation, it was necessary to distinguish the two types of excimers clearly, so Oyama et al. used two kinds of solutions. One contained 99% of untagged PEO and only 1% PEO*, where PEO* refers to PEO whose chain ends are tagged by pyrene. In this solution, PEO* is believed to behave like an individual PEO chain, which provides only intramolecular excimer. The other contained fully tagged PEO, which provides both intramolecular and intermolecular excimers.

Figure 1a,b shows the changes in Ie/Im (the intensity ratio of excimer to monomer) with addition of a PAA solution to the former solution containing PEO(4800) or PEO(9200). We see that the initial value of Ie/Im is higher in Fig. 1a than in Fig. 1b. This difference is due to the easier aggregation of terminal pyrene groups of the shorter PEO chain as well as to the larger diffusivity of the shorter PEO chain. The addition of PAA induced a marked decrease in Ie/Im of PEO*, which is greater with PAA (890 K) than with PAA(1850). The intramolecular mobility of the PEO chain is suppressed by addition of PAA, and this renders intramolecular excimer formation difficult. The interaction is stronger for the longer PAA as indicated by the larger decrease in Ie/Im.

These findings are further supported by the data on *inter*molecular excimer formation shown in Fig. 2, which were obtained by subtracting the data in Fig. 1 from the corresponding data for the latter solutions. Figure 2a shows that Ie/Im for PAA (890 K) initially increases, passes a maximum and levels off at [PAA]/[PEO]= 3. On the other hand, Ie/Im for PAA(1850) increases more slowly to a plateau at [PAA]/[PEO]≈4. The initial Ie/Im increase in either case implies that the PEO chains start to aggregate sidewise with added PAA by hydrogen bonding. According to Fig. 2a,b, this intermolecular aggregation at low [PAA]/[PEO] is stronger for the higher molecular-weight PAA, but its dependence on the molecular weight of PAA becomes weak at high [PAA]/[PEO]. Besides, it appears that the complexation is facilitated with the increase in the polymer concentration. Importantly, these results from fluorescence studies do not support the previously reported conclusion [3,24] that PAA/PEO forms complexes with a 1:1 base ratio at which the two complementary components come to perfect match.

Oyama et al. [23] have pointed out that the fluorescence technique is more sensitive to the formation of intermolecular complexes than the conventional methods including potentiometry, viscometry and turbidimetry. The following reports substantiate this. Using the conventional techniques, Antipina et al. [16] found that PAA and PEO started interacting significantly only when the PEO molecular weight was increased to around 6000. Ikawa et al. [17] observed no viscosity changes for PAA/PEO when the PEO molecular weight was lower than

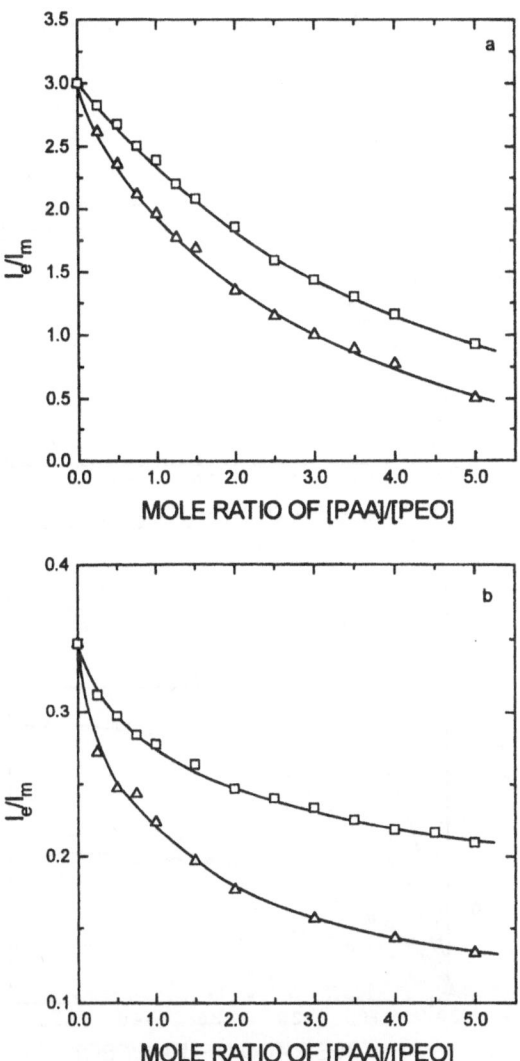

Fig. 1a,b. Changes in Ie/Im, the *intra*molecular excimer-to-monomer intensity ratio with the addition of PAA to PEO at 303 K. **a** Measured for the mixture of 1% PEO* (4800) and 99% PEO (4800), [PEO*+PEO]=1×10^{-3} M, [PAA]=1×10^{-1} M: ($\triangle$) PAA (890,000); ($\square$) PAA (1850); **b** Measured for the mixture of 1% PEO* (9200) and 99% PEO (9200) [PEO*+PEO]= 2×10^{-3} M, [PAA]=2×10^{-1} M: ($\triangle$) PAA (890,000); ($\square$) PAA (1850) [23]

8800. However, the fluorescence technique was able to detect complexation between PAA and PEO at a PEO molecular weight as low as 4800. It is worth noting that Chen and Morawetz [13] and Bednár et al. [15], who used dansyl-labeled PAA, observed only a small change in fluorescence intensity at PEO molecular weight of 8000 and no change at 3400.

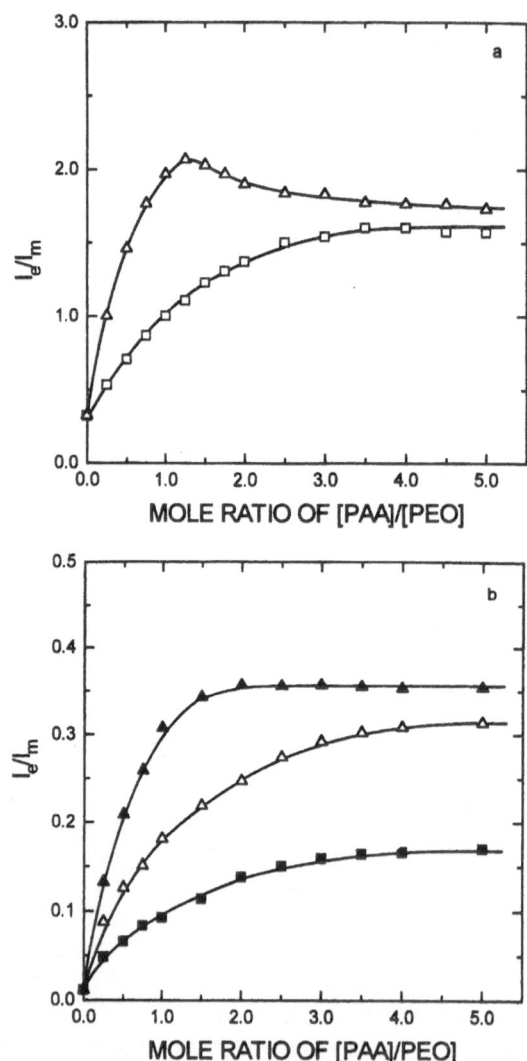

Fig. 2a,b. Changes in Ie/Im, the *inter*molecular excimer-to-monomer intensity ratio with the addition of PAA to PEO* at 303 K. **a** [PEO*(4800)]=1×10⁻³ M: (Δ) [PAA(890 K)]=1×10⁻¹ M; (□) [PAA(1850)]=1×10⁻¹ M; **b** PEO*(9200), (▲) [PEO*]=2×10⁻³ M, [PAA(890,000)]= 2×10⁻¹ M; (Δ) [PEO*]=1×10⁻³ M, [PAA(890,000)]=1×10⁻¹ M; (■) [PEO*]=2×10⁻³ M, [PAA(1850)]=2×10⁻¹ M [23]

From a study using pH, density, calorimetric and cloud point measurements, Eagland et al. [25] also cast doubt on the previously suggested 1:1 stoichiometric ratio for complexation and concluded that the critical molar mass of PEO for the complexation with PMAA was as low as 200. They found that the molar mass of PEO and the state of the resultant PEO solution play crucial roles in the mechanism of complex formation. When the PEO molar mass is 2000, complexation occurs between already pre-existing dimers of oxyethylene units (OE) with an MAA unit via hydrogen bonding. When the PEO molar mass is higher than 2000, the pair interaction between OE groups is minimal and single OE units make hydrophobic contact with MAA followed by hydrogen bonding of another OE or MAA, depending on the relative concentrations of PEO and PMAA. This mechanism for the complexation of PEO with PMAA is more complex than those previously reported by Oyama et al. [23] and Tsuchida and Abe [2].

2.3
Effect of Structural Factors on Complexation

2.3.1
Hydrophobic Interaction in PMAA/PEO Complexes

The secondary binding forces leading to complexation include Coulombic, hydrogen-bonding, van der Waals, charge transfer and hydrophobic interactions. The last is different from the others because the hydrophobic interaction is caused by rearrangement of water molecules rather than by a direct cohesive force between the molecules. Owing to the presence of 2-methyl groups, the conformation of PMAA, its coil structure in water, and its complexation with proton-accepting polymers receive strong hydrophobic effects. This problem has been investigated in comparison with PAA by many authors [26–28]. PMAA is more tightly coiled than PAA and the long-range attractive interactions between its hydrophobic α-methyl groups more strongly oppose the expansion of charged PMAA [26]. The local compact structure formed in uncharged PMAA in aqueous solution was found to be stabilized by hydrophobic interaction [27]. Although hydrogen bonding is a significant driving force for complexation the intermolecular complex PMAA/PEO is stabilized by hydrophobic interaction as evidenced by a smaller minimal chain length and a larger critical pH for complexation of PMMA than that of PAA [3].

Frank et al. [29] studied the effect of hydrophobic interaction by comparing the fluorescent properties of PMAA/PEO* and with those of PAA/PEO*. Here PEO* denotes pyrene end-labeled PEO. Figure 3 shows the intensity ratio Ie/Im of *intra*molecular excimer pyrene for PMAA/PEO*(9200) and PAA/PEO*(9200). It is seen that when added, PMAA more markedly reduces intramolecular excimer formation in PEO than does PAA. This difference is thought to be due to a stronger ability of PMAA to combine PEO* and the consequent suppression of intramolecular cyclization of PEO.

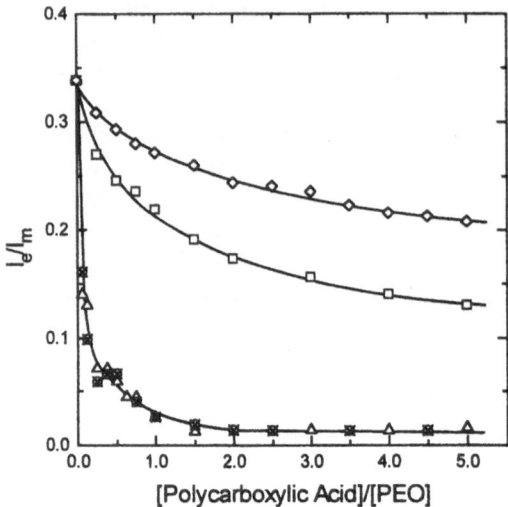

Fig. 3. Comparison of *intra*molecular excimer formation in PMAA/PEO* (9200) and PAA/PEO* (9200) at 303 K: (◇) PEO*+PAA(1850); (□) PEO*+PAA (890,000); (△) PEO*+ PMAA (9500); and (⊗) PEO*+PMAA (1530) [29]

Figure 4 compares the changes in Ie/Im of *inter*molecular excimer pyrene in PMAA/PEO*(9200) and PAA/PEO*(9200). The increase in intermolecular excimer emission is much greater for PMAA than for PAA at low [PMAA]/[PEO], even when the molecular weight of PMAA is lower than that of PAA, which suggests that complexation of PEO* with PMAA causes much larger pyrene accumulation in the complex than does that with PAA. Ie/Im for PMMA/PEO* passes through a maximum at a composition much lower than the stoichiometric value and decreases continuously as [PMAA]/[PEO] increases. This decrease can be attributed to the more compact and rigid structure of the complex, as is supported by an analysis of the peak position of the excimer emission.

The formation of a compact structure accompanying the complexation of PMMA with PEO and the lower flexibility of the PMMA chain in the complex than that of the PAA chain have been confirmed by viscometry [16], membrane contraction [2], and polarized luminescence techniques [3]. In addition, comparison of the dynamic light-scattering behavior of PMAA/PEO and PMAA/PEO* in solution shows that the pyrene label, which acts as a hydrophobic species, allows the labeled PEO to aggregate intermolecularly much faster than unlabeled PEO does [30].

Working on the complexation of PAA [31] and PMAA [32–34] with PEO monosubstituted with a variety of hydrophobic groups, Baranovsky et al. [31] found that the groups attached to PEO strongly enhanced the stability of both PAA/PEO and PMAA/PEO complexes, and pointed out that when complexation is examined using a fluorescently labeled component, the stabilizing effect of the hydrophobic fluorescent probes on complexation has to be fully taken into account for the interpretation of experimental results.

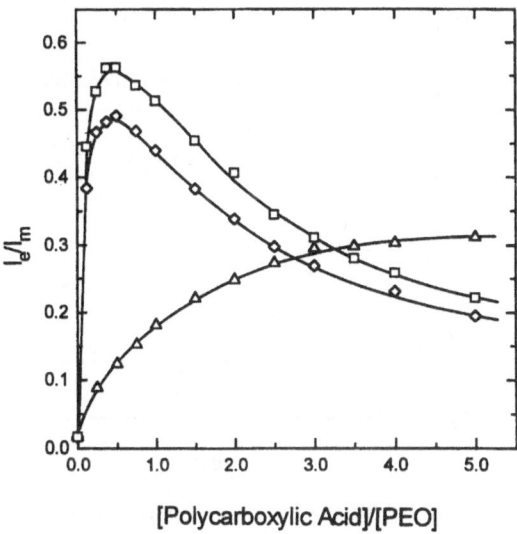

[Polycarboxylic Acid]/[PEO]

Fig. 4. Comparison of *inter*molecular excimer formation in PMAA/PEO* (9200) and PAA/PEO* (9200) at 303 K: (△) PEO*+PAA(890,000); (□) PEO*+PMAA (9500); (◇) PEO*+ PMAA (1530) [29]

2.3.2
Effect of the Degree of Hydrolysis of Polyacrylamide on Its Complexation with Proton-Accepting Polymers

PAM is one of the most extensively investigated water-soluble polymers. Studies on its complexation with PAA and poly(vinylpyrrolidone) started early [2,3]. By controlling hydrolysis and adjusting pH, it is possible to change the relative amounts of amide groups, undissociated and dissociated carboxylate groups in the modified PAM, which is denoted below as HPAM. Thus mixing HPAM with a counterpart polymer allows us to handle a broad range of interactions varying in type, strength and density.

Complexation of HPAM with PVPo, PEO [35] and PAA [36,37] has been extensively studied by fluorospectroscopy, with Ie/Im of the pyrene attached to HPAM used for estimating chain conformation. Ie/Im reflects the statistical conformation of a labeled polymer chain in such a way that a large value suggests greater polymer chain contraction and a small value greater polymer expansion and/or rigidity [35,36]. The changes in Ie/Im with pH for HPAM hydrolyzed to different degrees and for their blends with PVPo were examined, and the typical results obtained are shown in Figs. 5 and 6. For HPAM of low hydrolysis extent (EH) (Fig. 5), it is observed that its conformation is sensitive to pH. At low pH, COOH groups stay undissociated and the chain takes a coiled conformation leading to high Ie/Im. As the pH is increased, more acid groups dissociate and

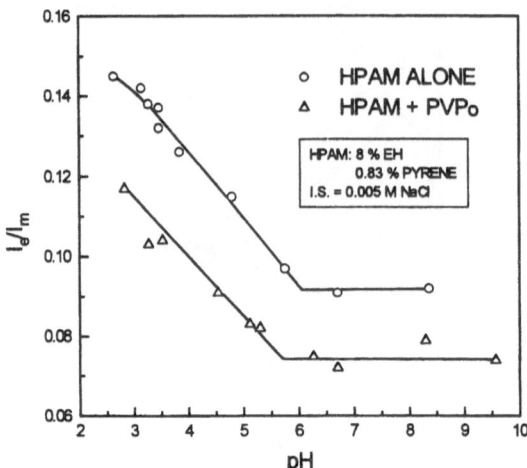

Fig. 5. pH dependence of Ie/Im for 8% hydrolyzed polyacrylamide and its mixture with PVPo [35]

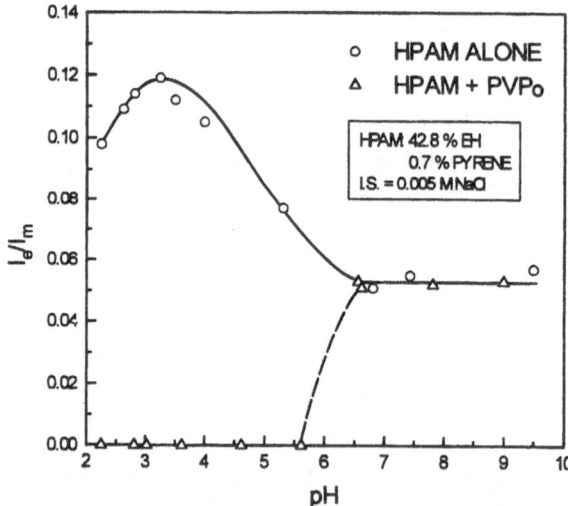

Fig. 6. pH dependence of Ie/Im for 42.8% hydrolyzed polyacrylamide and its mixture with PVPo [35]

repulse each other to make the chain more expanded and Ie/Im lower. However, for HPAM of high EH (Fig. 6), Ie/Im increases in the low pH range, which may be attributed to intramolecular bonding between the amide and acid groups. Mixing of PVPo with HPAM in solution, no matter how large EH may be, causes Ie/Im to vary appreciably. However, the pH dependence of Ie/Im varies with EH.

At low EH, Ie/Im is lowered to a nearly constant value over the entire pH range in the presence of PVPo. This behavior may be attributed to the relatively weak interaction between amide and pyrrolidone, both of which are non-ionic. In contrast, at high EH, Ie/Im is almost suppressed to zero at pH lower than about 5.8, which indicates that highly hydrolyzed HPAM forms a rigid and compact complex with PVPo. This remarkable Ie/Im suppression abruptly disappears when the pH is increased to 6, at which most of the acid groups are supposed to dissociate. In conclusion, the system of PVPo and HPAM should involve two kinds of interaction, one between amide groups and VPo which is relatively weak and pH-independent, dominates only when EH is very low, and leads to a loose complex, and the other between acid groups and VPo which is strong and pH-dependent, becomes dominant when EH is high and pH is low, and gives rise to a compact complex.

The same technique as above was used to study the interactions between HPAM and PEO [35]. The presence of PEO showed no detectable effects on Ie/Im of HPEM over the whole pH range. This finding indicates that the PAM unit is a very weak proton donor, not capable of forming a complex with PEO, in agreement with earlier reports [2]. It also indicates that the interaction between PEO and HPAM hydrolyzed up to 42.8 mol% is not strong enough to form a complex. Probably, because the acrylic acid units and acrylamide units distribute randomly along the HPAM chain [38], the sequence of acrylic acid units is not long enough to interact cooperatively with the ether units of PEO.

The pyrene-labeled hydrolyzed PAM was also used to elucidate its interaction with PAA by fluorospectroscopy [36,37]. The most remarkable feature of this system is the strong pH-dependence of the interaction. An increase of EH will reduce the number of possible amide/acid bonds between HPAM and PAA chains and enhance the intramolecular hydrogen bonding between the amide and acid groups. Therefore, for HPAM of high EH, complexation should occur only at low pH. In fact, it was found experimentally that as EH increased, the pH at which the interpolymer interaction began to occur shifted systematically to lower values. Two extreme cases are as follows. One is the complexation between PAM and PAA at pH lower than 5.0 and the other no complexation between PAA and HPAM of highest EH (42.8 mol%) over the entire range of pH.

2.3.3
Effect of the Degree of Ionization of Poly(carboxylic Acid) on Its Complexation with PEO

It is well known that poly(carboxylic acid) forms a hydrogen bond complex with a proton-accepting polymer as well as a polyelectrolyte complex containing a polycation. Partial ionization converts some carboxyl groups to dissociated carboxylate groups. The former is capable of hydrogen-bonding complexation but the latter not, so that partial ionization of PAA and PMAA provides a convenient way to adjust the proportion of active and inactive sites for complexation. Studies during the 1970s showed that there was a critical state of dissociation, above which no hydrogen-bonding complex was formed. Thus, the presence of a certain number of undissociated carboxyl groups is essential for PMAA and PAA to form stable complexes with PEO [3].

Iliopoulos et al. [22] studied complexation of proton-accepting polymers including PEO and PVPo with a series of PAA neutralized to different degrees, and found that even a low content of carboxylate sites (less than 15%) in the PAA chain was sufficient to prevent the chain from complexation. They emphasized the importance of a sufficiently long acid sequence for complex formation; in other words, complexation can occur only when the polyacid is neutralized below a certain ionization degree α_m characteristic of the system. Based on the results from viscosity and fluorescent polarization measurements on systems composed of variously ionized PAA and PEO or PVPo, Iliopoulos et al. [22] tried to explain the different structure characteristics of the complex as a function of the ionization degree α in the following way. (1) When α is close to zero and the polymer concentration is low, the acid group sequences are long enough so that the complex has a compact structure and the viscosity of the solution is low; and (2) when $0<\alpha<\alpha_m$ and, in particular, when α is close to $\alpha_m/2$, and the polymer concentration and molecular weight are sufficiently high, the complexes form a crosslinked structure. In fact, in the latter case, the PAA chain consists of long complexable sequences and short, uncomplexable sequences, so that it can interact with several polybase chains to give a gel-like structure. To mention more, by using the fluorescence polarization technique, Heyward and Ghiggino [39] found that only slight neutralization (<10%) of PAA was sufficient to break its complex with PEO.

Oyama et al. [40] treated the complexation of PMAA with PEO by using the same method as that for PAA/PEO [23]. The resulting Ie/Im for a series of PEO and PMAA with different degrees of ionization showed that the decrease in intramolecular excimer formation accompanying complexation diminished with increasing ionization. It is apparent that the presence of carboxylate groups in PMAA reduces the local PEO concentration in the complex, thereby suppressing excimer formation between pyrene groups on different PEO chains. However, the data for both intramolecular and intermolecular excimer formation indicated that the hydrogen-bond interaction between PMAA and PEO persists up to 30% ionization but no longer at 40%. This finding is in agreement with that of Anufrieva et al., who used fluorescence polarity measurements [41]. If compared with the results for the PAA/PEO system where complexation disappears at ionization degrees greater than 10–15% [22,39], it seems clear that PMAA/PEO has a much stronger ability to form a stable complex and tolerates a much larger content of the inactive sites for complex formation, because, in this system, hydrogen bonding is coupled with strong hydrophobic effects.

In addition to ionized acrylate groups being introduced into PAA chains by partial neutralization, other kinds of "structure defects", such as sulfonated groups and non-ionic groups, e.g. isopropylamide [-CONHCH$(CH_3)_2$] and hydroxyethyl (-CH$_2$CH$_2$OH), were incorporated into PAA by copolymerization or condensation reactions, and their influence on complex formation was examined. Viscometry and potentiometry studies [42] on the complex formation of PEO with PAA-copolymers containing carboxylate, sulfonate or isopropylamide groups revealed that under the conditions of low and constant ionic strength and constant pH, a significant difference appeared in the association behavior between the AA copolymers containing charged and neutral groups. No interpolymer complexation took place when the degree of substitution (DS) of ionic groups in the polyacid chain exceeded 10%, but the presence of non-ionic and relatively hydrophobic groups up to DS=30% allowed interpolymer association to occur. These results may be explained by assuming that the inactive charged groups not only interrupt the associating COOH sequences but also make the polymer chain less flexible owing to intrachain electrostatic repulsion and make it more difficult to find the PEO chains for interpolymer association. On the contrary, the presence of non-associating neutral groups in PAA may not have much influence on chain flexibility and enhances interpolymer association by hydrophobic interaction.

Krupers et al. [43] observed that structure defects introduced by non-ionic hydroxyethyl methacrylate (HEMA) into poly(acrylic acid) decreased the degree of complexation of carboxylic acid groups with PEO and lowered the efficient packing of PEO on the acrylic acid copolymer chains. When the AA/HEMA ratio of the copolymer exceeded 1:1, no complex formation was detectable.

2.4
Cooperative Effects in Complexation

One of the most important concepts in looking at polymer complexation is the cooperativity of the interaction between proton-donating and proton-accepting polymers. The existence of a minimum chain length needed for polymer association is generally accepted [18,21,22,39,44]. As mentioned in Sect. 2.3.3, only a small amount of inactive sites introduced in PAA may destroy its complexation with PEO or PVPo. It is preferable to avoid the complexities that, in partially ionized PAA, owing to electrostatic repulsions, the uncomplexable acrylate groups do not necessarily distribute randomly along the chain and their positions in the chain may change during complexation. Iliopoulos and Andebert [45], therefore, treated the complexation of a polybase (PVPo or PEO) with an acrylic acid based copolymer containing sulfonic acid groups, which are disassociated and uncomplexable, and observed a dramatic influence of the content of the sulfonic acid groups on the complexation. In some cases, even a low content of sulfonate groups prevented complex formation. These results are in good agreement with the prediction from a theoretical model incorporating the cooperativeness of the complexation [44]. However, the study by Wang and Morawetz [46] on the complexation of PAA with copolymers of acrylamide (AM) and N,N-dimethylacrylamide (DAAm) led to a different conclusion. They assumed that under a suitable pH condition (pH=4) only DAAm can be complexed by PAA while AM remains uncomplexable. Using the fluorescent probe technique they found that at pH=4 even the copolymer containing DAAm groups as low as 34 mol% underwent weak complexation. Obviously it contains only short complexable sequences. Based on these experimental results and considering the possible steric hindrances due to some local structure of the complexed sequences, Wang and Morawetz [46] concluded that the formation of stable hydrogen-bonded interpolymer complexes in aqueous solution involves no interaction of long sequences of monomer residues. In addition, Yang et al. [47], on the basis of their data on the complexation between phenol-formaldehyde resin and poly(N,N-dimethyl acrylamide), which will be discussed in Sect. 3, suggested reexamination of the need for many contiguous interaction groups to achieve stable complexation.

Iliopoulos and Audebert [45] explained the above-mentioned results in a different way. They reasoned that DAAm is a strong proton acceptor comparable to PVPo, so that its critical chain length must be small (~13 units) and the probability of finding such a sequence of DAAm in the copolymer is low though not negligible. With respect to the steric hindrance due to the local structure of the complexed sequence, they suggested that rapid rearrangement of hydrogen bonds caused some of them to break instantaneously and the system to stabilize dynamically.

In a series of investigations on polymer pairs having no inherent specific interactions, Jiang's group found that introducing hydroxyl groups, which in some cases was as small as less than 10 mol%, into one of the components, allowed them to complex with proton-accepting polymers in inert solvents. Obviously, in these systems with such low hydroxyl content, hydrogen bonds are separated by long sequences of inactive units, and such a sparse distribution of hydrogen bonds does not meet the requirement that a long sequence of uninterrupted hydrogen bonds be needed for complexation. The details of the work by this group will be presented in Sects. 5 and 6.

3
Interpolymer Complexes from Non-aqueous Media

We now move to a discussion on interpolymer complexation in organic media. The distinction from the previous section is only a matter of convenience. In fact, some interpolymer complexation may take place in organic solvents as well as in water. The investigation on complexation in non-aqueous media is relatively new, receiving special attention in the last decade and allowing significant information about the complexes in the solid state to be obtained. In the majority of the related reports, precipitates are regarded as complexes and solvent-cast products from homogeneous solutions as ordinary miscible blends. This distinction has some practical value but should not be carried too far. In fact, complexes, especially those formed with hydrogen bonds only in a small part of the segments in dilute solutions, may exist stably in solution, while the free-polymer component may co-precipitate with the complex.

3.1
Poly(vinyl Phenol) (PVPh) Used as a Proton-Donating Polymer

Among the proton-donating polymers used to study the complexation in non-aqueous media, poly(vinyl phenol) has been the most extensively used. Pearce's group [48] took up the complexation of PVPh with a series of proton-accepting polymers including poly(N,N-dimethyl acrylamide) (PDMA), poly(ethyl oxazole) (PEOX) and poly(vinyl pyrrolidone) (PVPo). Depending on the solvent used, the mixture may either precipitate one regarded as a complex. or give a clear solution which can be used to obtain solvent-cast 'blend' film (Table 1). Such solvent effect is ascribable to the difference in the hydrogen-bond-accepting ability of the solvent, which can be interpreted by the enthalpy-frequency shift relationship [49,50]. The relative strength of the solvent participating as the acceptor in the hydrogen-bonding interaction with PVPh can be evaluated by measuring infrared frequency shifts of the hydroxyl stretching absorption of phenol (a model of PVPh), which are 350, 340, 235, and 211 cm^{-1} in dimethylsulfone (DMSO), N,N-dimethylformamide (DMF), dioxane and acetone, respectively. The shift of the hydroxyl frequency is 325 cm^{-1} when phenol is mixed with N,N-dimethylacetamide (DMAc), a model compound of PDMA. This value is smaller than those of phenol in DMF and DMSO, the most strong acceptors, so that precipitation due to interpolymer interactions cannot take place in these solvents. Among the solvents used, methanol was the only proton-donating type, but was found to be a weaker hydrogen-bond donor than PVPh as the hydroxyl frequency shift in methanol/DMAc was 160 cm^{-1} smaller than 325 cm^{-1} in phenol/DMAc. Thus PVPh formed complex precipitates with both PDMA and PVPo in methanol. In all these pairs, the glass transition temperatures T_g of the complex precipitates were higher than those of the corresponding solvent-cast blends at the same compositions. This difference can be attributed to an extensive degree of interpolymer hydrogen-bond formation.

The effect of annealing on the T_g has been extensively studied. Thermal treatment raised the T_g of the "blends" as well as the complexes in PVPh/PDMA. This result can be considered as due to an increase in the number of hydrogen bonds caused by enhanced chain mobility accompanying annealing, which makes it easy for hydroxyl and amide groups to approach juxtaposition. However, this idea has not enjoyed support by FTIR studies which showed for PVPh/PDMA blends that the peak area of bonded carbonyl absorption relative to that of the total absorption decreased when the temperature increased above 143 °C. Therefore it was suggested that the increase in T_g accompanying annealing arises from improved chain packing, but no evidence has been obtained for the expected increase in macroscopic density.

The precipitates of PVPh/PDMA and PVPh/PVPo have T_gs higher than the weight averages of the T_gs of the component polymer. For example, the T_g of 50:50 PVPh/PVPo is about 210 °C, while the expected weight average is 150 °C. However, not all systems show the same trend; for example, the T_gs of PEOX/PVPh precipitates are always close but lower than the weight averages.

The precipitates of PVPh/PDMA from methanol and acetone solutions were examined by CPMAS ^{13}C NMR [51], and evidence for specific interaction was obtained with a 3 ppm shift in the phenolic carbon resonance peak. The proton spin-lattice relaxation times T_1 were shorter than those predicted by a linear model, though the rotating frame spin-lattice relaxation times $T_{1\rho}$ of the com-

Table 1. Precipitation behavior of blends of proton-donating poly(vinyl phenol) and various proton-donating polymers in different solvents

Proton acceptor	Solvent	Precipitation	Ref
PDMA	acetone	yes	[48,51]
$-CH_2-CH-$ $C=O$ N H_3C CH_3	dioxane	yes	
	methanol	yes	
	DMF	no	
	DMSO	no	
PVPo $-CH_2-CH-$ N $=O$	methanol	yes	[48]
	DMSO	no	
PEOX $-CH_2-N-CH_2-$ $O=C-C_2H_5$	dioxane	yes	[48]
	DMF	no	
PSVBDEP-13.3[a] $-CH_2-CH-$ $P=O$ CH_3CH_2O OCH_2CH_3	THF	yes	[54]
P4Vpy $-CH_2-CH-$ N	ethanol	yes	[58]
	DMF	no	
P2Vpy $-CH_2-CH-$ N	ethanol	yes	[58]
	DMF	no	
ST2VPy-70[a] $-CH_2-CH-CH_2-CH-$ N	ethanol	yes	[58]
	DMF	no	
PMPMA CH_3 $-CH_2-C-$ $C=O$ O N CH_3	ethanol	yes	[59]
	DMF	no	

[a] The numbers denote the molar fraction of interacting sites in the copolymer.

plexes were consistent with the prediction. The scale of homogeneity estimated from $T_{1\rho}$ data was about 2.5 nm.

Although many phosphonate compounds were reported to have strong proton-accepting strength [52], not much attention has been paid to phosphorus-containing polymers. Sun and Cabasso [53] reported evidence for hydrogen bonding between the hydroxyl group of cellulose acetate and the phosphoryl group of poly(4-vinyl benzenephosphonic acid diethyl ester). Zhuang et al. [54] studied cases where only part of the polymer segments contained interacting sites, using a series of styrene copolymers containing 4-vinyl benzenephosphonic acid diethyl ester (PSVBDEP) as a hydrogen-bonding acceptor. Blends of these copolymers with PVPh were examined by DSC, FTIR and NMR, with the results showing that PSVBDEP-4.3 (the numerals denote the molar fraction of interacting sites in the copolymer.) was immiscible with PVPh, but PVPh/PSVBDEP-7.5 had a single T_g when the concentration of the copolymer was higher than 60 wt%. Mixing of PSVBDEP-13.3 with PVPh caused both to precipite in THF. The IR spectra gave 354 cm^{-1} for the frequency shift of the hydroxyl group of phenol in the mixture of the PVPh with the VBDEP monomer. This value is comparable to the shifts reported for phenol in DMSO and DMF.

The pyridine unit is known as a strong proton acceptor. Although the miscibility of polymer blends composed of pyridine-containing polymers has been extensively studied [55–57], the complexation of these systems has received little attention. Studying the complexation of PVPh with such pyridine-containing polymers as poly(4-vinyl pyridine) (P4VPy), poly(2-vinyl pyridine) (P2VPy) and poly(styrene-co-2-vinyl pyridine) (ST2VPy) containing 70% pyridine units, Dai et al. [58] found that these polymers all formed complex precipitates in ethanol but not in DMF. The T_gs of the interpolymer complexes were substantially higher than the weight averages. P4VPy showed a stronger complexation ability than P2VPy. This is probably because the former suffers less steric hindrance than the latter in forming complexes. Recently, Jiang's group has made a systematic study on the complexation of copolymers containing pyridine units over a broad composition range with various proton-donating polymers. We defer a discussion of the results to Sects. 5 and 6.

Piperidine (pK_b=2.88) is a base stronger than imidazole (pK_b=7.05) and pyridine (pK_b=8.75). Therefore, polymers containing it are likely to form complexes with poly(vinyl phenol) by hydrogen-bond formation between piperidine and hydroxyl groups. However, according to Luo et al. [59], although poly(N-methyl-4-piperidinyl methacrylate) (PMPMA) complexed with PVPh in ethanol over the entire composition as judged by precipitation, evidence by FTIR indicated that the driving force for complexation would be the hydrogen bonding between the carbonyl of PMPMA and the hydroxyl of PVPh. Furthermore, X-ray photoelectron spectroscopy showed that the nitrogen atoms in the piperidine groups were not involved in intermolecular interactions with PVPh, as likely due to the steric effect of the N-methyl groups. Elucidating this steric effect needs the investigation of blends of polymers containing piperidine but no N-substituting group.

3.2
Phenol-Formaldehyde Resin Used as a Proton-Donating Polymer

The group of Pearce and Kwei [47,60] chose another phenolic hydroxyl-containing polymer poly[(1-hydroxy-2,6-phenylene)methylene] (PHMP) as a proton-donating polymer for studying hydrogen-bonding complexation, though the molecular weight of the sample was as low as about 1000. PDMA and PEOX were selected as proton-accepting counterparts. According to the results obtained, PHMP and PDMA form complex precipitates from acetone, dioxane and ethyl acetate solutions, despite the fact that these solvents are proton acceptors. This finding is understandable if the enthalpy–frequency shift relationship is taken into account [49,50]. The hydroxyl frequency shift of phenol in DMAc, a model compound of PDMA, is 340 cm^{-1}, which is definitely larger than those of phenol in the above solvents, i.e. 235, 218 and 187 cm^{-1} in acetone, dioxane and ethyl acetate, respectively.

In order to gain further insight into the critical length of the hydrogen-bonding sequence required for complex formation, Yang et al. [47] studied the complexation of PDMA with PHMP whose hydroxyl groups were partially converted to methyl ethers to reduce the hydrogen-bond density. Interestingly, precipitation occurred up to a conversion of about 40%, but not at higher conversions in dioxane. This finding implies that a much shorter uninterupted sequence than expected is sufficient to cause precipitation, and supports the argument put forward by Wang and Morawetz [46] (see Sect. 2.4) about the cooperative effect in complexation.

As to the T_g vs. composition relationship, PHMP/PDMA blend systems show positive deviations from the weight-average law and PHMP/PEOX blends exhibit sigmoidal curves. Thus, in the latter, T_g is higher or lower than the weight average depending on whether PEOX is the minor or major component. This behavior can be interpreted in terms of the competition of hydrogen bonding and mixing [61].

PHMP/PDMA blends in some mixed solvents do not precipitate but allow cast-film to be made. The T_gs of the "blend" films are always lower than those of the complexes and thermal treatment increases them. For example, T_g of the 50:50 PHMP/PDMA blend increased from 101 to 129 °C after 3 h annealing at 180 °C. Just as in the case of PVPh/PDMA discussed above [48], an intuitive explanation of this increase in T_g is that the extent of hydrogen bonding increases during the annealing process. In fact, for PHMP/PDMA blends the ratio A_b/A_t of the peak area of the hydrogen-bonded carbonyl to the area of the total absorption increases by annealing. However, this is not the case for the blends of PDMA and the modified PHMP. Annealing at 180 °C resulted in an increase in T_g but a decrease in A_b/A_t. Yang et al. [47] proposed that this behavior of A_b/A_t be viewed as a result of the competition between two opposite effects: the normal breaking of hydrogen bonding and the chain rearrangement by increased thermal agitation.

Chatterjee and Sethi [62] worked out the complexation of *p*-bromophenol-formaldehyde copolymer (PPF) with PVPo and PEO in the mixed solvent acetone/methanol (v/v 84:16), and found that PEO on the PPF chain was substituted with PVPo when the latter was added to the solution of PPF/PEO. This finding indicates that PVPo has a stronger ability for complexation than PEO, in conformity with the observations in aqueous solutions where PAA was used as the proton donor (see Sect. 2.2).

3.3
Proton-Donating Polymers Containing Aliphatic Hydroxyl

In order to see whether polymers containing aliphatic hydroxyl groups, which are proton donors weaker than the phenolic species, can also complex with polymers containing substituted amide groups, Dai et al. [63,64] investigated blends consisting of SAA copolymers of styrene/ally alcohol (AA) with 4.5 or 6.5 wt% AA contents and each of three isomeric amide polymers, PDMA, PEOX and poly(*N*-methyl-*N*-vinylacetamide) (PMVAc) (Table 2). These isomeric polymers have different abilities to form complexes with SAA. Thus, PEOX actually forms no complex with SAA in any of the solvents used. PDMA formed a complex in MEK but not in THF, and PMVAc complexes in both THF and MEK, so that the complexation ability is in the order of PMVAc>PDMA>PEOX, which is parallel with their hydrogen-bonding strengths estimitated by IR studies [63,64].

3.4
Poly(alkyl Itaconate)s Used as Proton-Donating Polymers

During the past decade, the properties and applications of the monoesters and diesters derived from itaconic acid have received growing interest because the monomers can be obtained biotechnologically and a great variety of related polymers can be synthesized thanks to two lateral esterifiable groups contained in the monomer [65]. The first work on poly(monomethyl itaconate) (PMMI)/PVPo complexes was done by Bimendina et al. [66], and it was followed by Pè-

Table 2. Complexation of SAA with some proton-accepting polymers

Proton donor	Proton acceptor	Solvent	Precipitation	Ref.
SAA-4.5	PDMA	MEK	yes[a]	[63]
		THF	no	
SAA6.5	PDMA	MEK	yes[b]	[63]
		DMF	no	
SAA4.5/6.5	PEOX	MEK, THF, DMF	no	[63]
SAA4.5/6.5	PMVAc	MEK	yes[c]	[64]
		THF	yes[d]	
		DMF	no	

[a] When PDMA<50 wt%.
[b] When PDMA<75 wt%.
[c] For entire composition.
[d] Only for SAA-rich cases.

rez-Dorado et al. [67,68], who reported on poly(monobenzyl itaconate) (PM-BI)/PVPo complexes. In recent years, Katime's group have systematically studied the complexation of polyitaconates with a series of proton-accepting polymers including PMMI/PDMA [69], PMMI/PEOX [69], PMMI/PVPy [70], PMMI/PVPo [71], poly(monoethyl itaconate) (PMEI)/PVPo [72], and PMBI/PVPo [73]. These polymer pairs showed the following characteristics in common.

Firstly, the formation or inhibition of intermolecular complexes, judged by precipitation in common solvents, strongly depends on the solvent. As shown in Table 3, they all form complexes in the weak proton-donating solvent methanol or some mixed solvents, but not in pure strong proton-accepting solvents such as DMF and DMSO. PMMI/PDMA, PMMI/PEOX and PMEI/PVPo all have a strong tendency to precipitate in 1:1 mole ratio from methanol.

Secondly, the IR spectrum of PMMI is rather complicated owing to the presence of (a) two kinds of carbonyl in the ester and acid groups; (b) self-association between the acid groups forming a dimer; (c) hydrogen bonding between hydroxyl and ester carbonyl, and (d) anhydrides formed during the polymerization process. Nevertheless, FTIR studies have confirmed that the dominant driving force for the complexation is hydrogen bonding between the proton-donating hydroxyl in the carboxyl groups of PMMI and the proton–acceptor groups in the counterpart polymers. Thus, for example, for PDMA and PEOX the amide I mode frequency, which mainly reflects the carbonyl stretching in amide groups, was 1642 and 1643 cm^{-1}, respectively. However, a new band at 1609 cm^{-1} associ-

Table 3. Complexation of poly(monomethyl itaconate) with various proton-accepting polymers

Proton acceptor	Solvent	Precipitation	Ref.
PV2Py, PV4Py	methanol methanol/DMF methanol/DMAC, etc.	yes	[70]
SPV4Py-50	methanol	yes	[70]
SPV4Py-25	methanol/THF (50:50)	yes	[70]
PDMA	methanol, ethyl glycol water, DMSO N-methylformamide (MF) DMF, DMAc	yes no no no	[69]
PEOX	water, methanol DMSO, DMF, MF, DMAc ethylene glycol	yes no no	[69]
PVPo	methanol	yes	[66,71]
PVPo[a]	methanol	yes	[67,68,73]
PVPo[b]	methanol, ethanol DMAc, dioxane pyridine, DMSO, DMF, MF	yes yes no	[72]

[a] Poly(monobenzyl itaconate) was used the proton-donating polymer.
[b] Poly(monoethyl itaconate) was used as the proton-donating polymer.

ated with hydrogen-bonded carbonyl appeared in both complexes [69]. The relative amounts of carbonyl groups associated and non-associated in the complexes can be determined by using spectral curve fitting methods. For the complex of PEMI with PVPo [72], a new band located at 1639 cm^{-1} was found, which can be associated with the hydrogen-bonded carbonyl in PVPo having a 41 cm^{-1} red shift relative to the free carbonyl.

Thirdly, PMMI starts thermally degrading at about 154 °C to form an anhydride [65,74]. Thermal analysis showed that the onset temperatures for degradation of the complexes PMMI/PDMA, PMMI/PEOX and PEMI/PVPo was substantially higher than those of the corresponding poly(monoalkyl itaconates), and the difference was attributed to the fact that hydrogen bonds in the complexes need to be broken before anhydride formation takes place.

In this series of studies on the polyitaconate-containing blends, too, the precipitates from solutions and the cast films from homogeneous solutions were named complexes and blends, respectively. In the case of PMMI/PDMA, complexes were obtained from methanol and blends from DMF. Interestingly, IR spectra of the complexes did not differ from those of blends having the same composition [69], but the two species had different properties, e.g. the complexes began degradation at temperatures higher than the blends did.

4
Miscibility Enhancement by Hydrogen Bonding

Progressive elucidation of the effect of specific interactions on the miscibility in polymer blends has been one of the most noticeable achievements in the study of polymer blends in the last two decades [75,76]. Many experimental results have supported the conclusion that the presence of specific interactions in a blend including hydrogen bonding, ion-ion pairing, etc. favors the enthalpy for mixing and allows the components to mix completely. In the earlier investigations, this conclusion was drawn from studies on the blends composed of polymers which contain interacting groups as an intrinsic part of the polymer structure. It has not only provided a practical guideline for making miscible blends from existing polymers but has also opened up a new area that aims to enhance miscibility in polymer blends by chemical modification or copolymerization which incorporates some chemical groups capable of creating specific interactions into one or both of the immiscible component polymers. In the early 1980s, several approaches to this new area were reported. Thus, Eisenberg's group [77–79] used cationic and anionic groups in immiscible polymer pairs, and Percec's group [7] reported the miscibility enhancement by interaction between pendant electron-donor and electron-acceptor groups. As for hydrogen-bonding interaction, the group of Pearce and Kwei [80,81] studied blends of modified polystyrene comprising hydroxyl-containing units and a counter-polymer containing carbonyl groups. Among all the interactions examined, hydrogen bonding seems most attractive, since it quite efficiently improved the miscibility without accompanying much change in the properties of the component polymers.

4.1
Theoretical Considerations and Miscibility Expectation

The group of Coleman and Painter [82] has combined an association model with the Flory–Huggins theory of polymer solutions to develop a theory describing miscibility behavior of polymer blends with hydrogen bonding, see also a review article [83]. Here we briefly describe their basic methodology and the conclusions which are useful in discussing the miscibility in the systems with controllable hydrogen bonding.

The theory is built on the major assumption that the free energy of mixing ΔGm is contributed by "weak" or "physical" interactions and "strong" or "chemical" interactions. Thus, ΔG_m for a pair of polymers A and B is given by:

$$\frac{\Delta G_m}{RT} = \frac{\phi_A}{M_A}\ln\phi_A + \frac{\phi_B}{M_B}\ln\phi_B + \chi\phi_A\phi_B + \frac{\Delta G_H}{RT} \tag{6}$$

The first two terms, where ϕ and M are the volume fraction and polymerization degree of A or B, respectively, represent the combinational entropy due to mixing. Since these terms are usually small, ΔGm is dominated by the balance between the third and fourth terms, i.e. $\chi\phi_A\phi_B$ and $\Delta G_H/RT$. The former refers to the physical forces unfavorable for mixing and is controlled by the Flory parameter χ, which may be estimated from solubility parameters for a set of carefully chosen groups which are free from association effects as much as possible. The latter comes from hydrogen bonding and is more difficult to estimate. Its magnitude depends on two major factors. One is the relative strength of self-association to inter-association. Simply speaking, if the strength of inter-association between the two dissimilar polymers is greater than that of self-association of either of them, miscibility is favored. The other is the density of specific interacting sites in the blend. It is expected that otherwise immiscible blends will be made miscible if this density is increased by incorporating interacting groups.

Hydrogen bonds are continually breaking and reforming by thermal agitation and, according to the association model, there exist instantaneous distributions of "free" monomers B_1, hydrogen-bonded dimers B_2 and multimers B_h, which obey the reaction scheme:

$$B_1 + B_1 \xleftrightarrow{K_2} B_2 \tag{7}$$

$$B_h + B_1 \xleftrightarrow{K_B} B_{h+1} (h > 1) \tag{8}$$

where K_2 and K_B are the equilibrium constants for the self-associations of proton-donating species B.

The inter-association of the proton-donating and proton-accepting groups can be described by:

$$B_h + A \xleftrightarrow{K_A} B_h A \tag{9}$$

where K_A is the equilbrium constant for the inter-association.

In the methodology of Coleman et al., K_2 and K_B for polymers containing hydroxyls are assumed to have the same values as those calculated from the FTIR data on appropriate model compounds. In general, K_A can be derived directly from IR studies of the single-phase blends. With the values of K_2, K_B and K_A known in this way, ΔG_H can in principle be calculated and then the phase behavior of the blend becomes predictable. The book by Coleman et al. [82] illustrates a wide range of hydrogen-bond-containing blends whose calculated phase diagrams agree well with observed results. Here we take up the cases where vinyl phenol (VPh) and p-(1,1,1,3,3,3-hexafluoro-2-hydroxy propyl)styrene (HFS) are used as proton-donating units.

From Table 4, we see that although the K_A values for these hydroxyl-containing units with the carbonyl or methacrylate are comparable, the self-association ability of PVPh is much stronger than that of HFS (note the differences in the K_2 and K_B values). The weaker self-association ability of HFS is probably due to the steric hindrance of bulky hexafluoroisopropyl groups. The relative strength of inter-association to self-association, as measured by the ratio K_A/K_B, may be taken as the relative efficiency that a particular hydroxyl group renders a polymer blend miscible. The difference in K_A/K_B between HFS and VPh gives rise to distinctly different miscibility behavior as shown in Figs. 7a,b. These miscibility maps predict that poly(alkyl methacrylates)s (PAMAs) with up to 3 methylene groups (PBMA) are miscible with PVPh and those up to 35 methylene groups with PHFS. In practice, it is difficult to verify experimentally the prediction for blends containing PAMA with very long side chains, though it has been reported that PAMA with 17 methlene groups was miscible with STHFS-60 [83,84]. Another conclusion of particular importance for the subject of this review is that, for both STVPh and STHFS, only a few percent of the hydroxyl units introduced render the copolymers miscible with PAMAs, provided that the side chain is not very long. This conclusion agrees with a large body of experimental results, which will be discussed below. However, in the experimental studies carried out by Jiang's group, discussed mainly in Sects. 5 and 6, have shown that, in the single phase area, a progressive increase in the content of hydroxyl in the copoly-

Table 4. Dimensionless self- and inter-association equilibrium constants values determined from miscible polymer blends scaled to a common reference volume (100 cm^3/mol) [83]

Polymer	Self-association K_2, K_B	Inter-association with carbonyl in methacrylate K_A
$-CH_2CH-$ ⬡ ÓH	21.0, 66.8	37.8 [K_A/K_B=0.6]
$-CH_2CH-$ ⬡ CF$_3$CCF$_3$ ÓH	4.3, 5.8	21.8 [K_A/K_B= 4]

mers transforms ordinary miscible blends to complex blends, in which unlike segments tend to pair together. Nevertheless, no existing theory predicts such behavior.

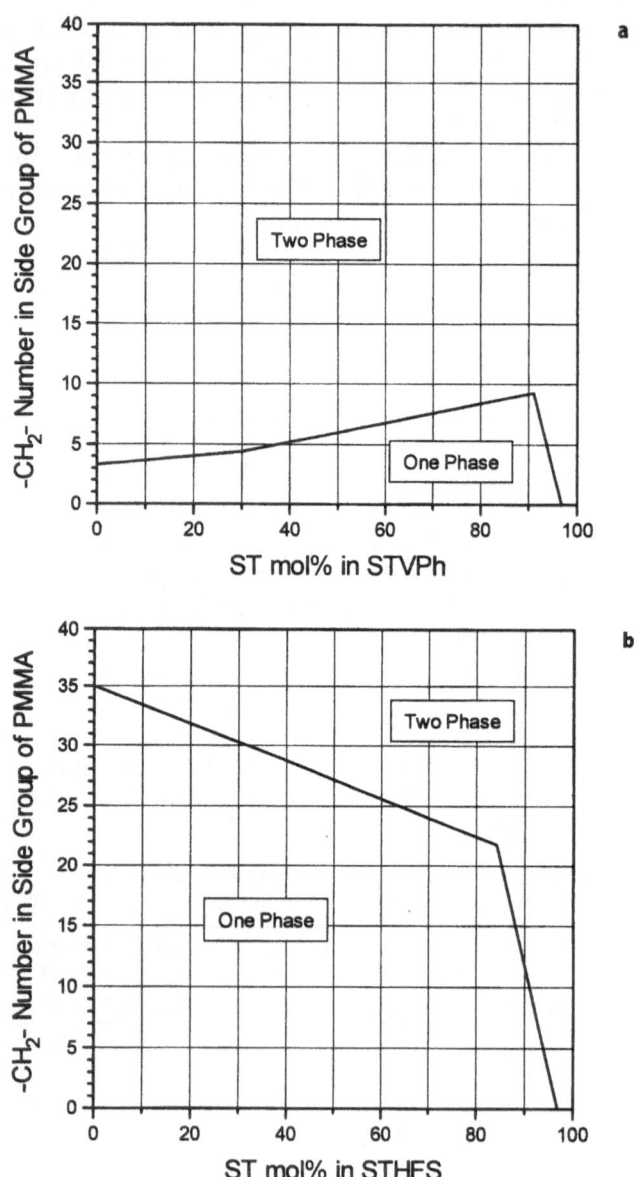

Fig. 7a,b. Miscibility maps calculated at 25 °C for poly(alkyl methacrylate) blended with **a** STVPh and **b** STHFS [83]

4.2
Miscibility Enhancement in Various Blends

A systematic study on miscibility enhancement by the introduction of hydrogen bonding has been made since the early 1980s by the group of Pearce and Kwei [85–87]. Most of the proton-donating polymers they used were copolymers of styrene and HFS, the latter being known as a strong hydrogen-bond donor [50,88]. These copolymers were blended with a variety of proton-accepting polymers including PMMA, PEMA, PBMA, PVAc, polycarbonate, amorphous polyamide and some polyesters. None of these polymers are miscible with polystyrene. The miscibility of the solvent-cast blends was judged by DSC measurements, with particular attention to determining the minimum content of hydroxyl-containing groups in the copolymers required to get a single phase for a given blend pair. The most remarkable result obtained was that only very small amounts of HFS incorporated in polystyrene were able to make the blends with PAMAs miscible. For example, 1.1, 1.8, and 3.9 mol% of HFS were sufficient to achieve miscibility over the whole composition range for PEMA, PBMA and PMMA, respectively. As for such counter-polymers having strong intrapolymer interactions as commercial polyester and polyamide, a hydroxyl content as high as 40 mol% was needed. The potential of HFS as a strong proton donor is further demonstrated by the fact that HFS-modified poly(dimethyl siloxane), which otherwise is immiscible with almost all polymers, can be made miscible with PBMA and PEO [86].

Cloud point curves measured for the blends of STHFS with PMMA, PEMA and PBMA, all had LCST, as expected for blends with specific interactions, and depended strongly on the degree of PS modification. For example, the lowest

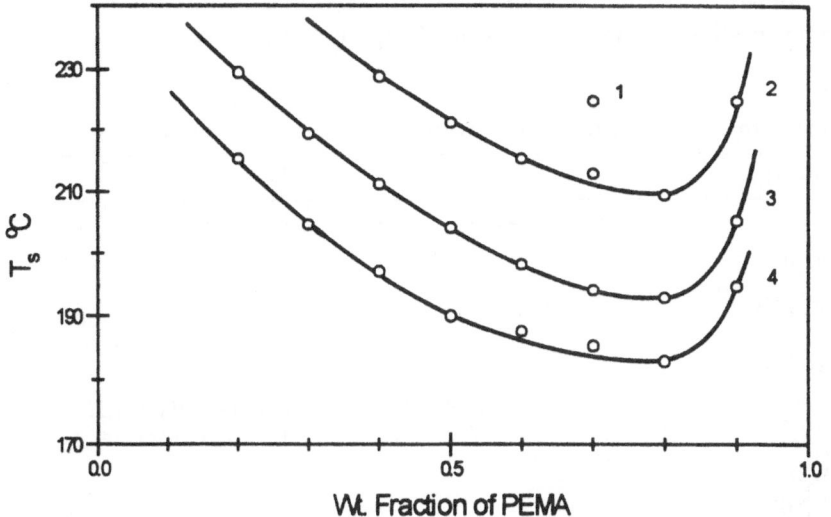

Fig. 8. Cloud points of STHFS/PEMA blends at a heating rate of 2 °C min⁻¹ *1* STHFS-1.5, *2* STHFS-1.3, *3* STHFS-1.2, *4* STHFS-1.1 [87]

cloud point of the blend STHFS/PBMA was 169, 188 and 207 °C for the hydroxyl contents of 1.8, 2.4 and 2.9 mol%, respectively, in the modified PS (Fig. 8).

The influence of the stereoregularity of PMMA on its miscibility has been studied using the blends with introduced hydrogen bonds. First, STVPh-5, i.e. polystyrene with only 5 mol% vinylphenol units, was used as the counterpart polymer, and it was shown by T_g and NMR spin-lattice relaxation time measurements that syndiotactic poly(methyl methacrylate) (s-PMMA) was miscible with this modified PS over the entire composition range. However, so was isotactic poly(methyl methacrylate) (i-PMMA) over only a narrow composition range, i.e. i-PMMA being larger than 70 wt% [89]. Further studies showed that a higher hydroxyl content, i.e. 10 mol% of HFS in STHFS, achieved miscibility over the whole composition range for both s-PMMA and i-PMMA [90]. The aggregation of PMMA segments became less by introduction of hydrogen bonding in both series of blends, but the T_g of STHFS-10/s-PMMA exhibited a positive deviation from the weight-average law and that of STHFS-10/i-PMMA a slight negative deviation. In addition, FTIR studies revealed that the s-PMMA blends slightly favor hydrogen bonding between hydroxyl and carbonyl groups over the i-PMMA blends.

Long-term research in Jiang's group on miscibility enhancement by introduced hydrogen bonding was motivated by the encouraging results of Pearce and Kwei, who showed the potential of the proton-donating group HFS for controlling miscibility of polymer blends. Instead of HFS, Jiang et al. chose α-methyl styrene based comonomer p-(1,1,1,3,3,3-hexafluoro-2-hydroxypropyl)-α-methyl styrene (HFMS) for their studies (Table 5) [91–93]. Considering that free-radical homopolymerization of α-methyl styrene is difficult because of its low ceiling temperature [94], we believe that HFMS groups are rather uniformly distributed along the copolymer chains. PS(OH) and PI(OH), which denote HFMS-modified polystyrene and polyisoprene, respectively, were prepared by

Table 5. Proton-donating units and the corresponding polymers in Sects. 4–6

Structure	Polymers	Ref
 F₃C-C-CF₃ OH (HFS)	STHFS	[82–87,95,107]
 CH₃ F₃C-C-CF₃ OH (HFMS)	PS(OH), PI(OH)	[84,91–93,108,111,112,121–125,135,136,141–145,150]

Table 5. Proton-donating units and the corresponding polymers in Sects. 4–6 (cont.)

Structure	Polymers	Ref
H‐C‐CH$_3$ / OH (HES)	PS(s-OH)	[141,142,158,159]
H$_3$C‐C‐CH$_3$ / OH (HPS)	PS(t-OH)	[141,142,148,158]
COOH	CPS	[82,141,142,149,158,160]
SO$_3$H	SPS	[107,113,114]
CH$_3$ / H‐C‐CH$_3$ / OH	MPαMS	[96–98]
CH$_3$ / H‐C‐CF$_3$ / OH	MPαMS	[96–98]
OH (VPh)	STVPh	[82–84,89,107,141,142, 146,147, 152,153,157]
H$_3$C‐Si‐CH$_3$ / OH (VPDMS)	copolymer of styrene and VPDMS	[100–103]

Table 5. Proton-donating units and the corresponding polymers in Sects. 4–6 (cont.)

Structure	Polymers	Ref
(N-maleimide structure)	poly(styrene-co-N-male-imide)	[99]
(N-maleimide)		
$F_3C-\overset{\underset{|}{OH}}{\underset{|}{C}}-CF_3$	PSe-OH	[91,92]

Table 6. Frequency of OH stretching in the H-bond between PS(OH)-9, 7 and the counter polymers, and the frequency shifts related to free hydroxyl groups. Blend compositions: PS(OH)/counter polymer 50:50 (w/w) [91]

Counter polymer	PMMA	PEMA	PBMA	PVA	PBA	PC	PSF
Frequency (cm^{-1})	3400	3398	3400	3370	3399	3440	3406
Shift (cm^{-1})	200	202	200	230	201	160	194

free-radical polymerization, and PSe-OH, i.e. polystyrene with $(CF_3)_2(OH)C$ end groups, was produced by anionic polymerization followed by termination with hexafluoroacetone.

The IR spectra of the blends which consist of either PS(OH) or PSe-OH as one component and one of a series of proton-accepting polymers as the counterpart component have remarkable features in common. These are distinct decreases in band intensity at 3600 and 3520 cm^{-1}, characteristic of the free and self-bonded hydroxyl groups, and the emergence of a new band at a lower frequency, giving evidence (Table 6) for the presence of strong interaction between the hydroxyl groups and the proton-accepting groups in the counter polymers.

Table 6 presents the frequencies of the bands for hydroxyl groups forming hydrogen bonds with the counter polymers and the corresponding frequency shifts from the band for free hydroxyl groups. The shifts are almost equal to those reported by Pearce et al. [85] for the corresponding blends, in which the hydroxyl groups come from HFS. This fact indicates that the methyl substituent in HFMS has no effect on the strength of intermolecular hydrogen bonding. In addition, the IR spectra of all blends of PMMA with the HFMS monomer, PS(OH), PSe-OH and PI(OH) show the stretching bonded hydroxyl located at 3400±1 cm^{-1}, which means that the strength of the hydrogen bond is independent of whether the hydroxyl group is in the monomer or connected to polystyrene. Moreover, the shifts are the same for hydroxyl groups either randomly distributed along the polymer chain or located only at the end chain. Kwei et al. [95] compared the FTIR spectra of the blends of HFS-modified polystyrene and several polymers containing proton-accepting groups with those of the mixtures of hexafluoroisopropanol and the corresponding proton-accepting low molecular weight compounds, all in dilute solutions of CCl_4, and found that the shifts of the hydroxyl stretching frequency were independent of the length of the chain to which the functional groups were attached. This chain-length independence of the frequency shift observed by the two groups implies that the hydrogen-bond-forming ability of the $(CF_3)_2(OH)C-$ group is great enough to overcome the influences of other chemical environments.

The miscibility of blends composed of a poly(α-methyl styrene) based proton-donating polymer (MPαMS) and proton-accepting polymers were studied by Cowie's group [96,97], who incorporated each of two proton-donor units, i.e. methyl carbinol and trifluoromethyl carbinol (Table 5), into poly (α-methyl styrene). They found it possible to prepare single-phase blends with such proton-accepting polymers as PVME, PVAc, PVPo and PVPy if the concentration of the donor unit in the MPαMS was larger than 4 mol%. For all blends examined phase separation occurred when the temperature was raised. When poly(alkyl acrylate)s and poly(alkyl methacrylate)s were used as the counterparts, the blends containing trifluoro carbinol units exhibited LCST-type cloud point curves 20–40 K above those of the blends containing methyl carbinol units [98]. This difference indicates that the former is a stronger proton donor than the latter. Cowie and Reilly [98] compared the miscibility of the blends comprising poly(alkyl methacrylate) with different alkyl groups, and found that single -phase blends were obtained for methyl, ethyl and t-butyl derivatives and two-phase mixtures for n-propyl and n-butyl derivatives. The difference suggests that the average distance between donor and acceptor should be small enough for stable hydrogen-bond formation to occur. The n-propyl and n-butyl derivatives would be more extended than ethyl groups. The t-butyl derivative, though quite bulky, is equivalent to the ethyl unit with respect to the distance from the terminal methyl to the carbonyl unit so that it allows the donor polymer to approach closely. This idea is applicable to the blends comprising MPαMS and poly(dialkyl itaconate)s as well [98]. Besides the blends comprising dimethyl, diethyl and di-t-butyl derivatives, those comprising di-n-propyl or di-n-butyl derivatives also gave single-phase miscibility, though much higher hydroxyl contents in the MPαMS were needed. For example, for poly(di-n-butyl itaconate), the necessary concentrations of trifluoromethyl carbinol and methyl carbinol were as high as 53 and 94 mol%, respectively. Here the copolymers are actually no longer regarded as "modified polystyrenes".

In Sect. 4.1, we have discussed how the competition between self- and inter-association affects the miscibility of a proton-donating polymer with its counterpart. A good illustration of this effect is the experimental data presented by Vermeesch and Groeninckx [99] for blends composed of PMMA and polystyrene copolymers containing different amounts (8, 14, 21 and 45 wt%) of N-maleimide units (NMI). DSC measurements showed that the copolymers containing 8, 14 and 21 wt% of NMI were miscible with PMMA, but that containing the highest amount (45 wt%) of NMI was not. Undoubtedly, the driving force for the miscibility in these systems is the hydrogen bonding between the -NH group in N-maleimide and the carbonyl in PMMA. However, IR data indicate a strong

self-association between -NH and carbonyl in maleimide which causes the NH
stretching band to shift from 3440 to 3352 and 3249 cm^{-1} for the dimer and multimer, respectively. When blended with PMMA, the bonded -NH shifts to
3275 cm^{-1}, which means that the strength of inter-association is in between
those of dimer and multimer self-association. However, the blends become miscible owing to the change in the interaction from NH$\cdots$O=C (imide) to NH$\cdots$O=
C (MMA) that occurs when the concentration of ester carbonyl of PMMA increases above that of maleimide carbonyl (Fig. 9). In contrast, when the maleimide concentration is higher, as in NMI-45, self-association dominates over inter-
association owing to a considerable increase in the maleimide carbonyl concentration.

Another possible reason for the poor miscibility of the copolymer having the
highest maleimide groups with PMMA is the dispersion effect that the increased
content of maleimide in the copolymer makes the difference in solubility parameter between PMMA and the copolymer greater.

Competition between self- and inter-association for miscibility has also been
observed for silanol-containing polymers. In recent years, Lu et al. [100,101]
have developed a new convenient route for the synthesis of these polymers,
which utilizes direct oxidation of a polymer precursor containing an Si-H function with dioxirane. Strong self-association in the copolymers of styrene and 4-
vinyl-phenyldimethyl silanol (VPDMS) was made clear by a silanol stretching
shift as large as 303 cm^{-1}. It is even stronger than that in polymers containing
HFS and vinyl phenol. The blends of the styrene/VPDMS copolymer and PBMA
cast from methyl ethyl ketone showed a "miscibility window", i.e. miscibility occurred only for the copolymers containing VPDMS between 9 and 34 mol%. In
the blends, the hydroxyl stretching shift due to inter-association was only about
94 cm^{-1}. The fact that the copolymers with silanol contents higher than 34 mol%
are not able to form miscible blends with PBMA can be attributed to the domi-
nance of self-association over inter-association. The miscibility window for the
blends cast from toluene, a typical inert solvent for hydrogen bonding, was
found to shift to lower silanol contents, i.e. between 4 and 18 mol% [102, 103].
FTIR studies also revealed that mixing of the silanol copolymer with PVPo led

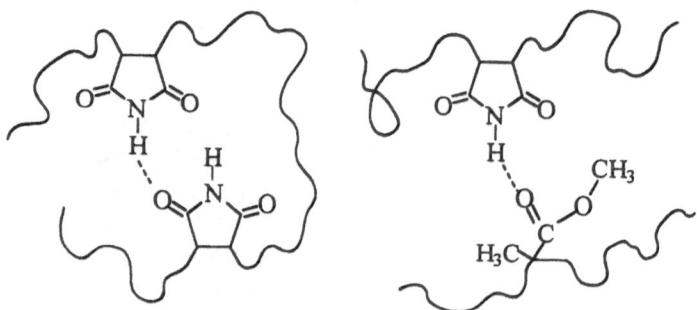

Fig. 9. Schematic representation of hydrogen bonding in styrene-co-NMI/PMMA blends

to strong inter-associated hydrogen bonding between the silanol and amide carbonyl groups, with a silanol stretching shift of 277 cm^{-1}. Thus, the blends not
only showed miscibility at silanol contents between 4 and 18 mol% but also precipitated at the silanol contents higher than 34 mol%.

4.3
Further Miscibility Study Using TEM and Fluorospectroscopy

Transmission electron microscopy (TEM) has been successfully applied to the
miscibility studies of polymer blends containing controllable hydrogen bonding. Jiang et al. [91,92] reported the TEM observations for two groups of binary
blends obtained by mixing PS(OH) with PMMA and PBA. Since no unsaturated
bonds existed in these polymer components, the routine staining agent OsO$_4$
was not applicable, and RuO$_4$ was used to stain the phenyl-containing component selectively [104].

A common conclusion drawn by Jiang et al. is the regular morphological variation with the hydroxyl content in PS(OH). To illustrate it the micrographs of
PS(OH)/PMMA (40:60), with PS(OH) having different hydroxyl content, are
shown in Fig. 10. Figure 10a for the polystyrene/PMMA blend containing no hydrogen bonding exhibits typical immiscible morphology, i.e. the PS phase as
large as 10–20 μm is dispersed in a PMMA matrix with distinct phase boundaries. The thin parallel stripes in the PS phase are believed to have arisen from the
"cutting effect" during microtoming rather than any finer structure. Incorporation of only 0.8 mol% of HFMS units in PS brings about an obvious morphological change, i.e. the average size of the dispersed phase diminishes by about one
order of magnitude (Fig. 10b). A more marked change occurs when the HFMS
content reaches 1.2 mol% (Fig. 10c). Although it is still possible here to distinguish the relatively bright and dark areas, heterogeneity appears only on a scale
as small as a few tenths of a nanometer and no clear phase boundary can be seen.
It is expected that the specific interactions enhancing miscibility become strong
enough to overcome the unfavorable interaction between the component polymers and make the blend completely miscible as more hydroxyl groups are incorporated. In fact, the blend PS(OH)-1.9/PMMA was transparent and showed
no phase structure under electron microscopy.

Ternary blends comprising PS(OH), PS and PMMA present complicated morphologies. A typical micrograph (Fig. 11a) of the blend in which PS(OH) contains 1.2 mol% hydroxyl content shows two kinds of dispersed phases in the
PMMA matrix, namely, a few large domains (3–5 μm) and many very fine domains (0.1 to 0.3 μm). Considering the fact that incorporation of hydroxyl
groups into polystyrene greatly improves its dispersion in the polyacrylate matrix, it is reasonable to attribute these fine and coarse phases to polystyrenes
with and without introduced hydroxyl groups, respectively. In fact, when more
hydroxyl groups are incorporated [e.g. PS(OH)-9.7], only the large dispersed
phase of PS is observed and the modified PS is completely dissolved in PMMA to
form a multicomponent matrix (Fig. 11b). It is significant that the complicated

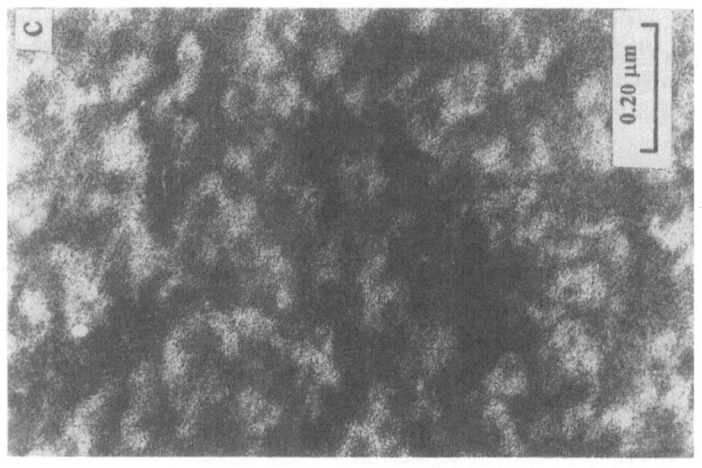

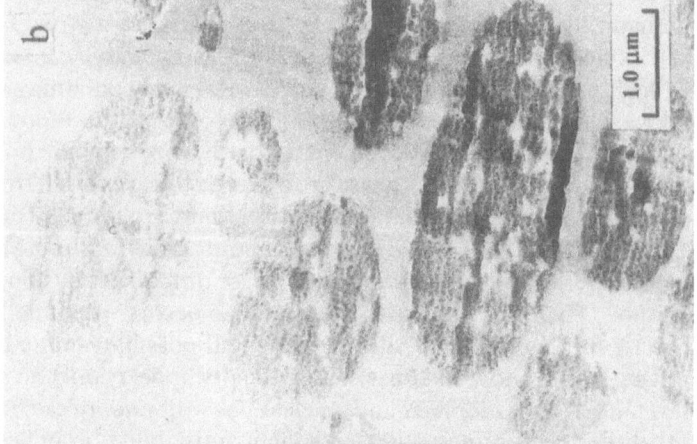

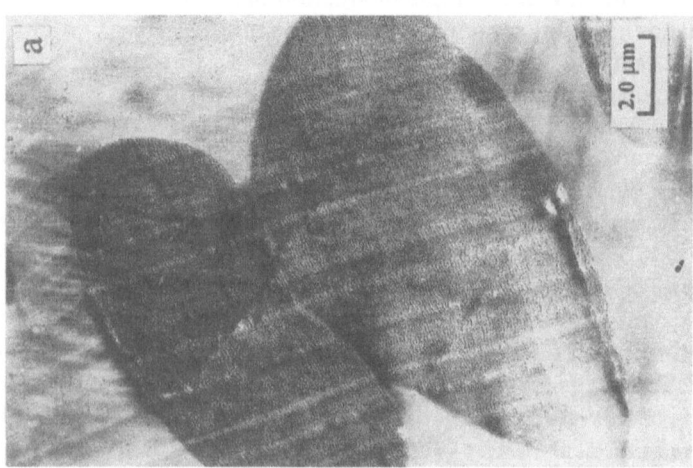

Fig. 10. TEM micrographs for PS(OH)/PMMA (40:60) blends with different hydroxyl contents in PS(OH). **a** 0 mol%, **b** 0.8 mol%, **c** 1.2 mol% [91]

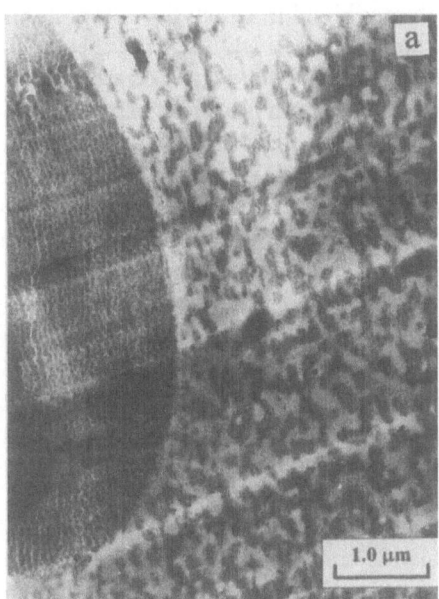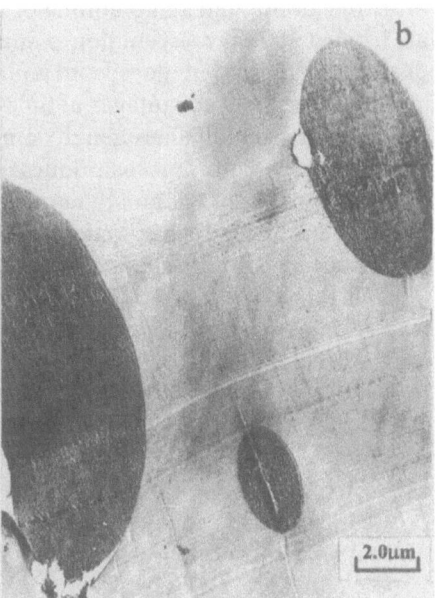

Fig. 11. TEM micrographs for ternary blends. a PS(OH)-1.2/PS/PMMA 1:1:3 and b PS(OH)-9.7/PS/PMMA 1:1:3 [92]

morphologies of these ternary blends can be understood if the miscibility of PMMA with PS(OH) by hydrogen bonding and the immiscibility of PMMA with PS by the unfavorable enthalpy of mixing are taken into account. This fact implies that introduction of hydrogen bonding into ternary blends makes their complicated morphological features controllable.

Morawetz et al. [105,106] were the first to use non-radiative energy transfer (NRET) fluorospectroscopy for exploring polymer-polymer miscibility. The basic principle is as follows. In a system containing two kinds of fluorescence chromophore, if the emission spectrum of one (donor D) overlaps the absorption spectrum of the other (acceptor A), a non-radiative energy transfer from the former to the latter may occur when the system is excited by irradiation that the former selectively absorbs. The efficiency of energy transfer (E) inversely proportional to I_d/I_a, where Id and Ia denote the emission intensities of D and A, respectively, depends on the average distance r between D and A according to the relationship:

$$E = R_o^6 / (R_o^6 + r^6) \tag{10}$$

Here, R_o is the distance at which half of the excitation energy is transferred and a constant for a given fluorescing pair in a given medium. E is sensitive to changes in r when r is comparable in order to R_o. Conventionally, instead of E, I_d/I_a proportional to its inverse is used as the variable for monitoring the proximity of a donor and an acceptor. For a polymer blend in which the components are labeled with fluorescent donor and acceptor, respectively, Id/Ia is expected to decrease as the system goes from a phase-separated state to a mixed state.

One of the great advantages of the NRET technique in miscibility studies is its sensitivity to spatial heterogeneity amounting to the order of 2 nm for usual fluorescing pairs. Thus, this technique can be an important supplement to existing techniques such as DSC and dynamic mechanical analysis which is limited to detecting heterogeneity on a scale of about 10 nm, though it provides only relative information about the change in the degree of mixing of the components.

Chen and Morawetz [107] applied NRET to the blends composed of carbazole-labeled PS containing different hydrogen-bond donors and anthracene-labeled poly(alkyl methacrylate). The results confirmed the study of Pearce [85] in regard to the role played by the hydroxyl groups in HFS, showing that the phenol group was almost comparable to HFS in the power of enhancing miscibility between polystyrene and polymethacrylate. In addition, it was shown that the PK_a value of the hydrogen-bond donor is not the only governing factor for miscibility.

Jiang's group has made efforts to explore the variations of energy transfer efficiency and morphology in a series of blends composed of carbazole-labeled PS(OH) and anthrancene-labeled PMMA as a function of the hydroxyl content in PS(OH), with the concentration of the chromophore labels kept at 5.2×10^{-3} M [108]. As shown in Fig. 12, with increasing the amount of hydroxyl groups introduced, the intensity ratio of carbazole to anthracene, Ic/Ia, first decreases monotonically and then suddenly levels off. The highest value of Ic/Ia for blend a, which contains no intercomponent hydrogen bonding, is typical of immiscible phases. The increase in the hydroxyl content enhances the energy transfer effi-

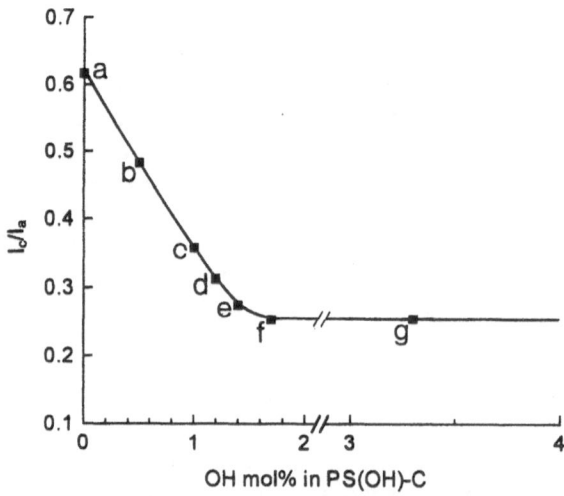

Fig. 12. Ic/Ia of blends PS(OH)/PMMA (50:50) as a function of hydroxyl content in PS(OH). Concentrations of carbazole and anthrancene labels are 5.17×10^{-3} M [108]

ciency because it makes the phase size smaller and the phase boundary more indistinct. When the hydroxyl content reaches 1.2 mol% (blend d), the two phases greatly interpenetrate each other and some domains shrink to only ~20 nm. A further increase in the hydroxyl content, as small as less than 1.4 mol%, brings about a substantial change in the morphology, i.e. the blend becomes transparent and TEM reveals no structure. Thus, the work by Jiang et al. has made clear that the efficiency of energy transfer in a blend varies with the introduction of hydroxyl groups and the variation reflects the transition of the blend from immiscibility to miscibility.

We note that all the films designed by e, f and g in Fig. 12 are transparent and exhibit no phase structure when examined by TEM, but differ somewhat in the degree of mixing when studied by the fluorescence technique. In fact, it was found that Ic/Ia for film e was slightly but definitely larger than those of the others. For the donor-acceptor pair used in this study, the characteristic distance is 2.7–2.9 nm [106], so that composition fluctuation on this scale in the films could be detected by routine TEM observations. The failure to see the phase structure in e, therefore, may be due to a great decrease in composition fluctuation by a strong intercomponent interaction.

4.4
Scale of Homogeneity of Mixing

We have mentioned above that the introduction of only 1–3 mol% of hydroxyl moiety to polystyrene allows the polymer to mix with poly(alkyl methacrylate)s. However, since the hydrogen bonds in the miscible blends are sparsely distributed and there are long sequences of segments between hydrogen bonds in both components, Pearce et al. [85] has put forward a question as to the scale of homogeneity of mixing. As is well known, the spin relaxation times T_1 and $T_{1\rho}$ in solid-state NMR measurements provide information about homogeneity on a scale of 20–30 nm and on a scale of 2–3 nm, respectively [109]. Thus, Pearce et al. [87,110] and Jiang et al. [111] used them to study blends containing vinyl phenol for the former, and hexafluoro carbinol for the latter as proton donors and obtained similar results.

All carbon resonances of pure PMMA, PS(OH), and their blends showed single-exponential decays in both T_1 and $T_{1\rho}$. The T_1 and $T_{1\rho}$ values for the main resonance lines are shown in Figs. 13 and 14, respectively. For each of PMMA and PS(OH) containing different amounts of hydroxyl, the protons attached to different carbons have similar relaxation times, which indicates that the spin diffusion equalizes the relaxation rates of all protons. In addition, the T_1 and $T_{1\rho}$ values for PS(OH) show a gradual decrease and increase, respectively, as the hydroxyl content in PS(OH) is increased. The difference in either T_1 or $T_{1\rho}$ between PMMA and PS(OH) is substantial, encouraging the gathering of information about the phase structure of their blends by NMR relaxation time analysis.

Figure 13 indicates for both blends PS(OH)-1/PMMA and PS(OH)-3 /PMMA that the T_1 values associated with PMMA and PS(OH) protons are different but

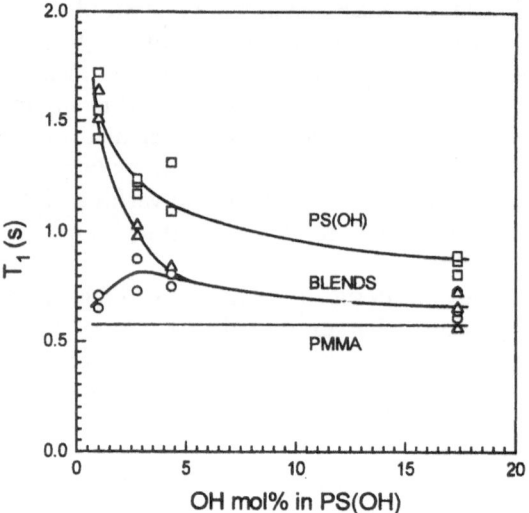

Fig. 13. Spin-lattice relaxation times T_1 of PS(OH) and PS(OH)/PMMA as functions of the hydroxyl content in PS(OH). □ for different protons attached to different carbons in pure PS(OH). Δ and ○ for different protons in PS(OH) and PMMA in the blends, respectively. *Thin horizontal line* for pure PMMA as a reference [111]

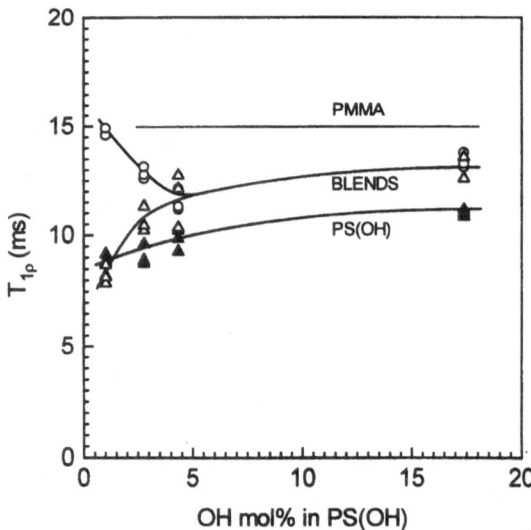

Fig. 14. Relaxation times in a rotating frame $T_{1\rho}$ for protons attached to different carbons in pure PS(OH) (▲), PS(OH) in the blends (△), and PMMA in the blends (○) as functions of the hydroxyl content in PS(OH). *Thin horizontal line* for pure PMMA as a reference [111]

relatively close to those of the pure components. This suggests that the blends have heterogeneity on a scale of 20–30 nm, but the situation changes considerably when the hydroxyl content in PS(OH) increases. Thus, in PS(OH)-4/PMMA, the T_1 of all protons of the components are almost the same, suggesting that increasing hydrogen-bonding density diminishes heterogeneity.

A similar situation is also obtained for $T_{1\rho}$ (Fig. 14). The values of $T_{1\rho}$ of PS(OH) and PMMA for blends with low hydroxyl content, such as PS(OH)-1/PMMA and PS(OH)-3/PMMA, are different, but the difference almost disappears for the blends with high hydroxyl content, such as PS(OH)-4/PMMA. This means that homogeneous mixing of the components takes place not only on a scale of 20–30 nm but also on a scale as low as 2–3 nm, i.e. miscibility is attained at a segmental level. In other words, the blends which are made miscible by the introduction of only a few hydrogen bonds are identical on the homogeneity scale to those whose components contain interacting sites in all of their segments.

On the basis of NMR data, the smallest OH content in PS(OH) capable of making the blends miscible is estimated to be between 2.8 and 4.4 mol%. It is larger than about 2 mol% by other techniques, and the discrepancy can be attributed to the difference in blend preparation. The film samples used in previous work were cast from toluene; a slow evaporation process favored interpolymer hydrogen-bonding formation. On the other hand, the powder samples used for NMR were precipitates obtained by adding a large amount of methanol to toluene or trichloromethane. Methanol, being a weak proton-donating solvent, probably impedes interpolymer hydrogen bonding so that a larger hydroxyl content in PS(OH) is required for miscibility.

It is interesting to note in Figs. 13 and 14 that the relaxation behavior of PS(OH)-18/PMMA and PS(OH)-4/PMMA is indistinguishable. However, as shown in the next section, these blends have quite different chain arrangements. This implies the limitation of routine NMR relaxation time measurements for monitoring the blend structure at the molecular level.

Jong et al. [110] studied the NMR relaxation of blends STVPh/PBMA (w/w, 1:1) with introduced vinyl phenol groups. The differences in both T_1 and $T_{1\rho}$ between STVPh and PBMA decreased as the hydroxyl content increased and finally disappeared as the content reached 4.4 mol%. Thus, it was concluded that the blends become homogenous on a scale smaller than 2.2 nm at this hydroxyl content. However, Campbell et al. [112] reported some different results from an NMR study on the blend of PS(OH)-1.5/PBMA (60:40). Though this blend was found to be homogeneous by light scattering at room temperature, their NMR data revealed compositional heterogeneity on a 0–12 nm scale remaining in it. Although 1.5 mol% HFMS may allow PS(OH) to mix with PBMA, this hydroxyl content is so close to the critical value that the degree of mixing of the component polymers would be very sensitive to the conditions for the preparation of the blend. No data for blends with higher hydroxyl content are available in the report.

4.5
Miscibility and Mechnical Properties

A lot of work has been done on enhancing miscibility of polymer blends by introducing hydrogen bonding, but little attention has been paid to investigating how the degree of mixing affects the macroscopic mechanical properties of the blends with hydrogen bonding. Register's group [113,114] studied this problem with two series of blends composed of a glassy component and a rubbery component, i.e. STVPh/poly(ethyl acrylate) (PEA) and partially sulfonated polystyrene (SPS)/PEA.

At low functionalization levels, both STVPh/PEA and SPS/PEA showed enhanced interfacial mixing as the hydroxyl or sulfonic group was introduced, when examined by DSC and dynamic mechanical analysis. However, at high functionalization levels, they exhibited miscibility changes of opposite trends. Thus, for STVPh/PEA, as the hydroxyl content increased, the glass transitions of the component polymers first gradually broadened, and as the hydroxyl content reached above 4.46 mol%, the transitions began narrowing and became more well-defined, indicating the approach to miscibility. On the other hand, for SPS/PEA, although clear evidence of miscibility enhancement was observed as the sulfonyl content was increased from zero to about 4.7 mol%, further increase of sulfonyl content brought about reappearance of the transition associated with pure SPS. This may be due to SPS aggregation. Interestingly, this miscibility difference between the two blends led to different relationships between the mac-

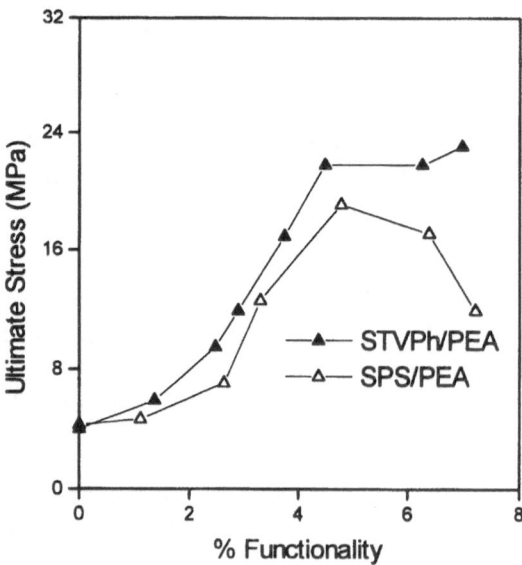

Fig. 15. Ultimate stresses for STVPh/PEA and SPS/PEA as functions of the functionality in HPS and EPS [113]

roscopic mechanical properties and the degree of functionality, as illustrated in Fig. 15, where the ultimate stress is plotted against the functionality for the two groups of blends. We see for both that as the functionality increases, the ultimate stress first increases gradually owing to improved adhesion of the polymer and then sharply increases reflecting their intensive mixing. However, with further increase in functionality, the ultimate stress for STVPh/PEA reaches a plateau and remains approximately constant, whereas the ultimate stress for SPS/PEA begins to decrease owing to component segregation. This difference in the mechanical response was interpreted by using the association model of Coleman et al. [82]. In fact, since the solubility parameters of polystyrene, poly(ethyl acrylate), poly(vinyl phenol) and poly(styrenesulfonic acid) are 9.20, 9.52, 10.89 and 15.35 $(cal/cm^3)^{1/2}$, respectively, SPS with introduced sulfonic acid has larger unfavorable dispersive interactions with PEA than STVPh does. At high levels of functionality, the dispersive interactions in SPS/PEA may overwhelm those from hydrogen bonding and induce phase separation leading to different mechanical properties.

4.6
IPNs with Intercomponent Hydrogen Bonding

The existing methods of improving compatibility or miscibility in polymer blends can be classified into two categories – physical and chemical. In connection with the former, interpenetrating polymer networks (IPNs) have been studied extensively [115,116]. In the process of preparing an IPN system, the presence of a crosslinked structure would retard phase separation, and enhance mixing to some degree, which is called 'enforced miscibility'. A considerable body of experimental evidence has led to the conclusion that the degree of mixing between the components in sequential IPNs gradually increases as the crosslink density of either the first or second network is increased.

As for the chemical approach, introducing specific interactions into blends is very practical and effective. Obviously, it is interesting to know how the phase structure, miscibility and properties of blends vary with simultaneous introduction of both intracomponent crosslinking and intercomponent hydrogen bonding. Nishi and Kotaka [117], studying IPNs formed with a complex-forming pair of polyethylene oxide (PEO) and poly(acrylic acid) (PAA), found that only IPNs with low crosslink density exhibited a single phase structure, but, as the crosslink density increased, PEO-rich and PAA-rich phases coexisted with the complex phase and finally existed exclusively. Thus, complexation between highly crosslinked PEO and PAA networks was not possible. Such disadvantageous effects of crosslinking on the miscibility was also reported by Bauer et al. [118] for systems of polystyrene and poly(vinyl ether). Some semi-IPN systems composed of miscible pairs were studied by Coleman et al. [82,119], and also by Kim et al. [120]. For such a system consisting of a miscible pair of phenol-formaldehyde resin and PMMA, it was observed that at high crosslinking temperature, the dissociation of hydrogen bonding was so extensive that only a small fraction of the original hydrogen bonding was recovered on cooling. This was probably because crosslinking hindered the attainment of equilibrium segmental interactions [120].

In order to elucidate the role played by crosslinking and specific interactions in determining the miscibility in IPNs, it is desirable to have IPN systems in which the densities of crosslinks and specific interaction are independently controllable over a broad range. The IPNs composed of poly(alkyl acrylates) and PS(OH), in which a strong proton donor containing $-C(CF_3)_2OH$ can be incorporated at will, meet this requirement. Jiang et al. [121,122] and Xiao et al. [123–125] prepared a series of full IPNs of PBA and PS(OH) by a typical sequential procedure and investigated with it the effect of hydrogen bonding at different levels of crosslink density and the effects of crosslinking at different levels of hy-

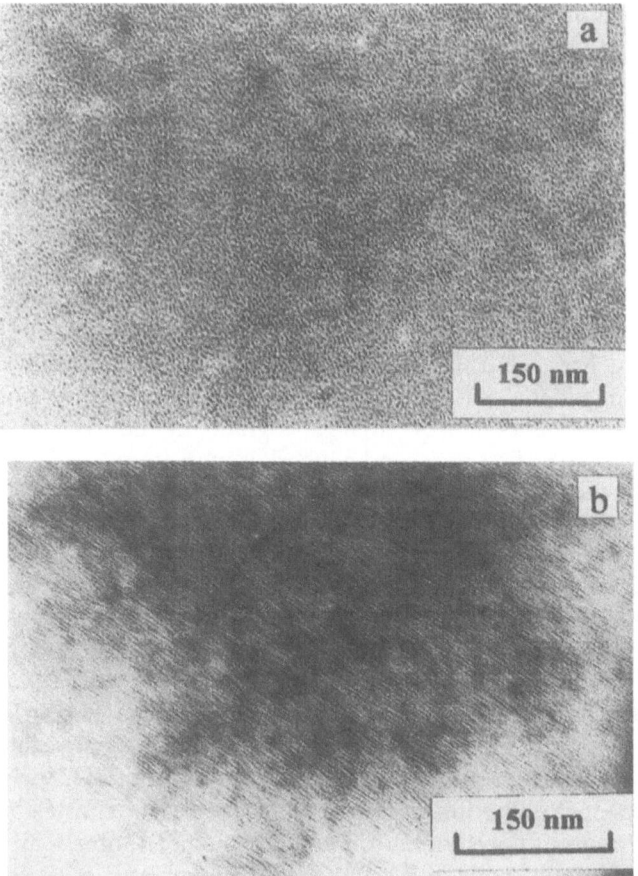

Fig. 16. TEM micrographs of PS(OH)-5/PBA (48:52) IPNs. The average molecular weight of PBA between crosslinks is 4200 and the crosslink agent content in PS(OH) network is **a** 1% and **b** 7% [123]

droxyl content on the morphology, miscibility and some other properties. The data from DSC, turbidimetry and TEM revealed that crosslinking favors miscibility for the IPNs with low hydroxyl contents (≤1 mol%) in PS(OH), while the corresponding uncrosslinked blends were immiscible. This advantageous effect of crosslinking undoubtedly can be attributed to the formation of an IPN which restricts molecular mobility and thus retards phase separation. However, for the PS(OH)s with high(5, 30 and 50 mol%) hydroxyl contents, though the corresponding uncrosslinked blends were miscible or even formed complexes (see Sects. 5 and 6), the IPNs showed two phases having extremely fine structures as illustrated in Fig. 16. Here, the phase structure is so fine and the phases so thoroughly interpenetrate each other that it is almost impossible to estimate the phase size at the resolution of TEM used. The causes for this disadvantageous crosslinking effect on miscibility on the segmental scale are complicated. A detailed study [123,124] with DSC and DMA showed that these IPNs possess inherent microheteregeneity owing to the presence of highly crosslinked microdomains of PBA, which may not be solubilized with PS(OH), whose hydroxyl content is as high as 30–50 mol%.

4.7
Homopolymer/Block Copolymer Blends with Controllable Hydrogen Bonding

The miscibility and morphology of AB/A-type blends composed of a block copolymer AB and the homopolymer A have aroused great interest since the early 1970s. Such blends may assume a variety of morphologies depending on the nature of phase separation, which are the microphase separation between unlike blocks A and B only, the macrophase separation between AB and A, and the coexistence of microphase and macrophase separation. Some great efforts in the past 20 years have elucidated some general effect on the morphology of the blends from the factors including the molecular weight of homopolymer A relative to that of the A block, the chain architecture of AB copolymer, the composition of the AB/A blend, the temperature and evaporation rate of solvent in casting [126]. In contrast to the many investigations with AB/A blends, AB/C-type blends have received little attention, where C and A are chemically different but may be miscible thanks to some specific interactions between them.

Tucker et al. [127,128] were concerned with systems composed of a homopolymer poly(2,6-dimethyl phenylene oxide) (PPO) and polystyrene-containing block copolymers such as diblock SI, triblock SBS, SIS and SEBS (here B and I refer to polybutadiene and polyisoprene, respectively, and SEBS to hydrogenated SBS, where the central block can actually be regarded as a random copolymer of ethylene and 1-butene). This selection of polymers for studying AB/C-type blends is obviously based on the miscibility of PS with PPO [129–131]. Here the specific interaction between the polymer chains is sufficient to overcome the unfavorable entropy of mixing the "bound" and "free" chains. The main findings are as follows: (1)The molecular weight of the PS block is a major factor determining the extent to which PPO/PS segments are mixed. (2) The solubilization of homopolymer PPO in the block domains of PS is obviously larger than that in AB/A blends. (3) Neither the molecular weight of PPO over the range examined ($2.4–39\times10^3$) nor the nature of the soft block and the location of the PS block significantly affect the miscibility of PPO in PS domains. (4) The observed broadening of the glass transition of the "hard phase" composed of PPO and PS blocks can be explained by the "gradient segment density model" proposed by Xie et al. [132–134]. For the study of AB/C type blends, aimed at elucidating the effect of specific interactions on the miscibility, it is desirable to have blend systems where the intensity of specific interactions between homopolymer C and block A is adjustable. The blends of a diblock copolymer composed of polyisoprene and poly(methyl methacrylate) blocks (PI-b-PMMA) and PS(OH) with different amounts of hydroxyl-containing units meet this requirement [135,136].

A series of blends composed of PI-b-PMMA and PS(OH)-1 with different molecular parameters cast from toluene showed the typical feature of combined macrophase and microphase separation, i.e. the immiscibility of PS(OH)-1 with PMMA blocks because of the low density of hydrogen bonding between them. However, the phase structure totally changed when the hydroxyl content in PS(OH) was increased to 1.6 mol%. Regardless of the composition and the molecular weight ratio of PS(OH)-1.6 to PMMA block (M_h/M_b), all the blends showed only microphase separation, which indicates miscibility of PMMA blocks with PS(OH). A typical example is shown in Fig. 17. In the whole area examined, spherical microdomains of PI are homogeneously distributed. The miscibility of PS(OH) with PMMA block causes the average distance between the PI microdomains to increase above that in the pure block copolymer. Differing from the AB/A blend, in which miscibility was only achieved when the molecular weight ratio M_h/M_b was smaller than 1 [126], M_h/M_b had very little effect on the miscibility for the AB/C blend. For example, miscibility between PS(OH)-1.6 and

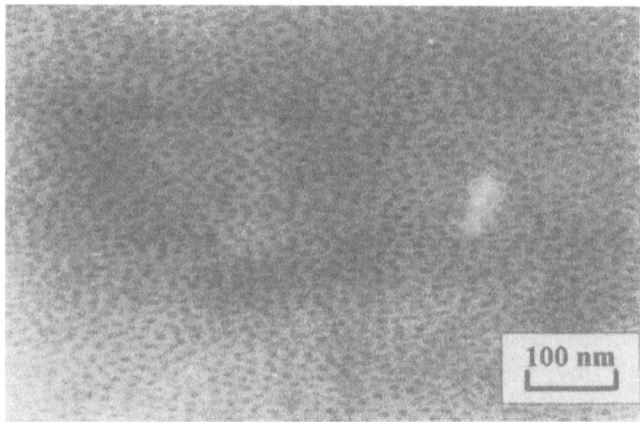

Fig. 17. TEM micrograph of PI-b-PMMA/PS(OH)-1.6 cast from toluene [135]

PMMA blocks was observed even when M_h/M_b was as high as 4.1. A further increase in the hydroxyl content to about 5 mol% in PS(OH) made the blend miscible even at larger M_h/M_b and high PS(OH) content.

It was generally difficult to examine the miscibility in PI-b-PMMA/PS(OH) by DSC because the difference in T_g between PMMA block and PS(OH) is very small. The difficulty can be solved by employing enthalpy relaxation [137–140]. When aged below the T_g, two separate enthalpy recovery peaks associated with PMMA blocks and PS(OH) appeared for blends with low (0 and 1.0 mol%) hydroxyl contents, but only a single peak for blends with higher (1.7 and 5.0 mol%) hydroxyl contents. This finding was in good accord with that from the TEM observations mentioned above. For the blend PI-b-PMMA/PS(OH)-1.7, a substantial broadening of the glass transition (ΔT_g=45 °C) was observed in the DSC thermograph. This phenomenon can be explained by the gradient segment-density model for block copolymer/homopolymer blends proposed by Xie et al. [132–134].

The above discussion is concerned with the blends cast from toluene, a typical inert solvent for hydrogen bonding. The study [135] on the blends cast from the proton-acceptor-type solvent THF gave quite different results. Thus, the blends composed of PI-b-PMMA and PS(OH) with different hydroxyl contents ranging from 1.6 to 5.0 mol% exhibited coupled macrophase and microphase separation, even when M_h/M_b was much smaller than 1. Figure 18 shows a typical micrograph for PI-b-PMMA/PS(OH)-5 taken at a relatively high magnification to see the interface between the block copolymer phase and the structure-free phase. The microdomain structure is quite similar to that of the pure copolymer, indicating rather complete phase separation of the block copolymer from PS(OH)-5. In fact, the homogeneous microphase structure characteristic of miscible blends appeared only when the hydroxyl content reached 10 mol%. This solvent effect on miscibility behavior is similar to that on complexation (see Sects. 3 and 5).

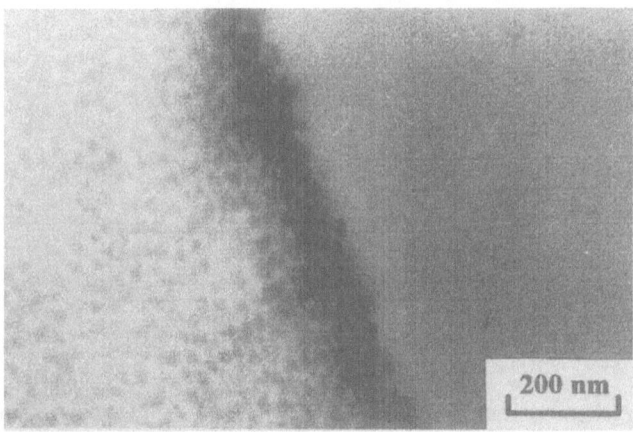

Fig. 18. TEM micrograph of PI-b-PMMA/PS(OH)-5 cast from THF [135]

5
Complexation of Blends with Controllable Hydrogen Bonding in Solutions

In Sects. 2 and 3 we referred to intermacromolecular complexation due to hydrogen bonding. The research on this subject began as early as the 1960s, and the polymers used in the early studies were usually limited to those which had either proton-donating or proton-accepting groups as their inherent parts. In Sect. 4 we discussed the miscibility enhancement of otherwise immiscible blends by introducing proton-donating groups into one of the components, thereby to induce interaction with the proton-accepting polymer. This subject began receiving attention in the late 1970s and the research on it has since become progressively active.

As shown schematically in Fig. 19, in the traditional complexation studies, every segment of each component often has a specific interaction group, while, in the studies on miscibility enhancement, only a small amount of interacting sites is introduced. However, the driving force for both complexation and miscibility is the same, namely, intermolecular hydrogen bonding. The question is whether it is possible to change an ordinary miscible blend to a complex blend by merely strengthening intermolecular hydrogen bonding. Jiang's group tried to answer this by systematically studying systems in which the density of intercomponent hydrogen bonding is adjustable [141,142], and observed that the

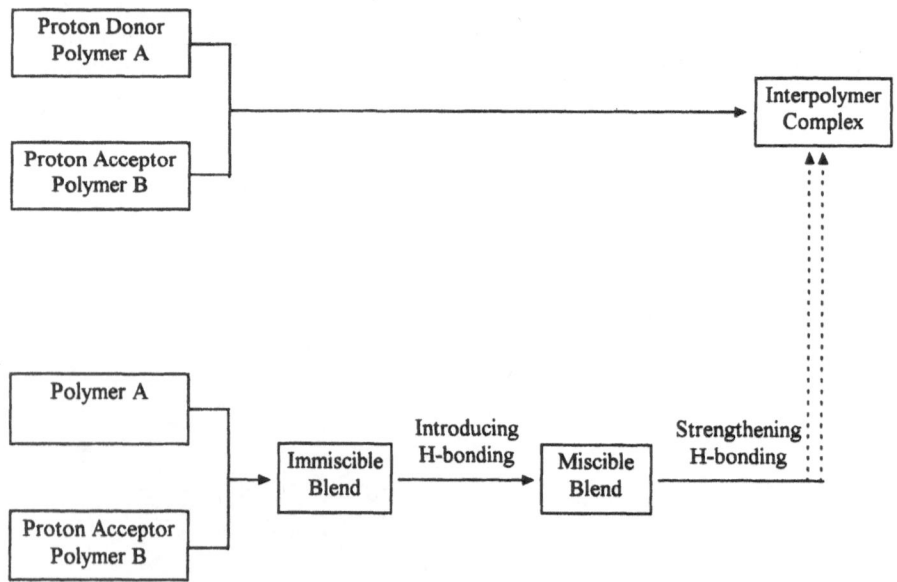

Fig. 19. Schematic representation of the relationships between hydrogen bonding and miscibility /complexation in systems containing inherent interacting sites and controllable hydrogen bonding

transition from separated polymer coils to complex aggregates took place in solution as the intermolecular hydrogen bonding was strengthened. Moreover, Jiang's group found that, in the solid state, further strengthening of hydrogen bonding transformed a miscible blend to the complex state. Complexation of polymer blends with controllable hydrogen bonding in solution is discussed in this section, and that in the solid state in the next section. The chemical structures of the proton-donor units mentioned in Sects. 4–6 are listed in Table 5.

5.1
Hydroxyl Content Dependence of Complexation [111,143–150]

Since polymer-polymer complexation in solution always accompanies contraction or collapse of the component polymer coils, viscometry has proved useful for detecting complexation. The composition dependence of reduced viscosity has been commonly used for estimating the composition of the complexes formed [2]. Jiang's group has paid special attention to the effect of hydroxyl content and hence hydrogen-bond density on complexation. Figure 20 shows the reduced viscosity (η_{sp}/c) of PS(OH)/PMMA in toluene as a function of the ratio of the total moles of styrene and HFMS monomer units in PS(OH) to the moles of PMMA [143]. The total concentration of 1.5×10^{-3} g/ml is far below the overlap concentration. It can be seen that, depending on the OH content in PS(OH), the

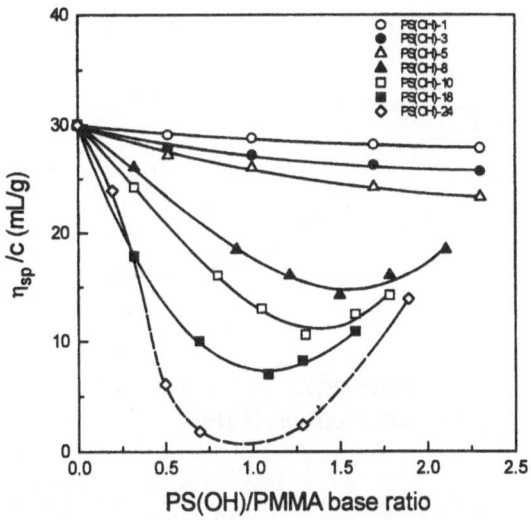

Fig. 20. Reduced viscosities for PS(OH)/PMMA in toluene as a function of the PS(OH)/PMMA base ratio. The original concentrations of the individual polymers are 1.5×10^{-3} g/ml^{-1}. The *dashed line* shows the data for PS(OH)-24/PMMA, taken after filtrating the precipitation. The hydroxyl contents in PS(OH)-1, 3, 5, 8, 10, 18 and 24 are 1.0, 2.8, 4.4, 7.9, 9.5, 17.4 and 24.0, respectively [143]

solutions show two kinds of viscosity-composition relationship. The viscosity for the mixtures containing PS(OH) with OH of 1, 3 and 5 mol% changes monotonically with the PS(OH)/PMMA base ratio, following rather closely the weight-average law. However, the viscosity for the mixtures with OH contents higher than 8 mol% first decreases sharply and passes through a minimum. This quite different viscosity behavior should reflect the difference in the assembly of the component chains. The regular variation of η_{sp}/c in the former may be taken to indicate that the polymer chains exist as separate coils, while the sharp decrease in η_{sp}/c in the latter may imply the collapse of the coils or their aggregation due to the formation of interchain complexes.

We mentioned in Sect. 4.3 that the efficiency of energy transfer between a fluorescence donor and acceptor depends strongly on the distance between the two species over the scale of ~2–4 nm. Since this scale is far smaller than the diameter of ordinary polymer coils in solution, NRET measurement may be expected to be sufficiently powerful to detect the chain interpenetration in solutions. Thus, it has been successfully used for the characterization of hydrophobic association in aqueous solution [151]. For exploring complexation an energy donor (c, carbazole) and an energy acceptor (a, anthracene) were incorporated into PS(OH) and PMMA, respectively [143]. The results are shown in Fig. 21, where the emission ratio of the donor to acceptor, I_c/I_a, is plotted against the hydroxyl content in PS(OH). Noting that the energy transfer efficiency is inversely pro-

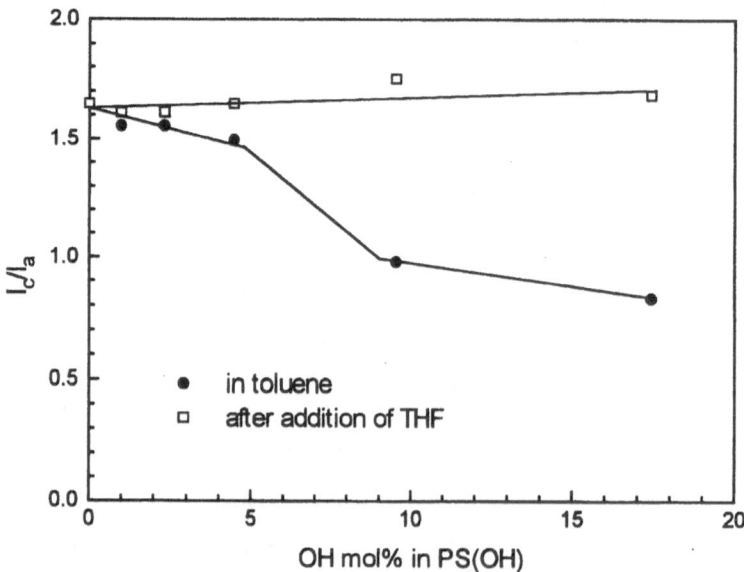

Fig. 21. □ Ic/Ia for carbazole-labeled PS(OH) and anthracene-labeled PMMA in toluene as a function of the hydroxyl content in PS(OH); ● corresponding data taken after a small amount of THF had been added [143]

portional to I_c/I_a, we see that the data can be classified into two levels. One is the low energy-transfer level, that is associated rather well with separated coils of the component polymers and observed at low hydroxyl content. The other is the high energy-transfer level, that reflects interpenetration of the unlike chains and is observed at high hydroxyl content. The transition from separate coils to complex aggregate is seen to occur in the range of OH content from 5 to 8 mol%, well consistent with the result from viscometry.

5.2
Direct View of Complex Aggregates by LLS [146–148,152–154]

Laser light scattering (LLS) has proved very useful for investigating aggregation processes especially in very dilute solutions, where conventional viscometry is often less sensitive. Although some LLS data was available on the aggregation of polymer pairs without specific interactions in non-aqueous media [155,156] and also polyelectrolyte pairs in water or in polar solvents [4,5], Xiang et al. [146,147] and Zhang et al. [152,153] were the first to use LLS for detecting interpolymer complexation in systems with controllable hydrogen bonding. Here, we refer to the study on the system STVPh/PEMA [157], in which vinyl phenol serves as a proton-donating group.

Figures 22, 23 and 24, respectively, depict the apparent hydrodynamic radius distributions $f(R_h)$ for STVPh-3, STVPh-9, and STVPh-15, as well as their blends with PEMA in toluene. Here, the STVPh unit fraction Fr is defined as the total moles of styrene and vinylphenol monomer units of STVPh relative to those of PEMA plus STVPh. In LLS experiments done, PMMA cannot be "seen", because its dn/dc in toluene is close to zero. As seen in Fig. 22, the $f(R_h)$ of the STVPh-3/PEMA (50:50, w/w) blend is similar to that of pure STVPh-3, indicating that

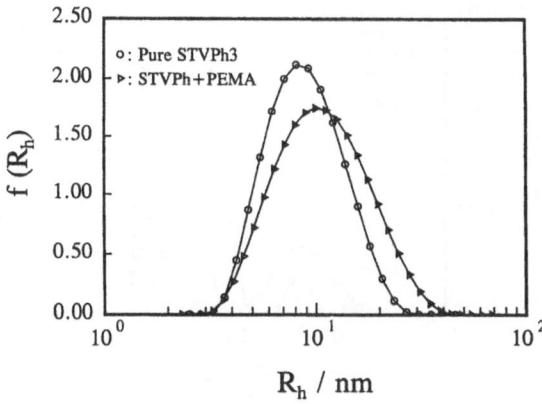

Fig. 22. Hydrodynamic radius distributions $f(R_h)$ for STVPh-3 and STVPh-3/PEMA blends (50:50, w/w) in toluene, determined at a total polymer concentration 1.0×10^{-4} g/ml and a scattering angle of 15° [147]

the presence of PEMA has no significant effect on the distribution of R_h. The $f(R_h)$ of pure STVPh-9 is rather broad, extending to R_h=100 nm (Fig. 23). This feature should arise from self-association of the hydroxyls in STVPh-9. The $f(R_h)$ of the blends STVPh-9/PEMA are quite different. The addition of only 3 wt% PEMA changes $f(R_h)$ to bimodal. The peak at a small R_h (ca. 20 nm) can be assigned to the free STVPh-9 and the peak at a large R_h (ca. 100 nm) to the STVPh-9/PEMA complex aggregates. Although the peak at 100 nm has a larger area than that at 20 nm, the weight fraction of the complex aggregates in the blend mixture is actually much smaller than that of the free STVPh, because LLS is more sensitive to larger molar mass particles. Further addition of PEMA causes the peak

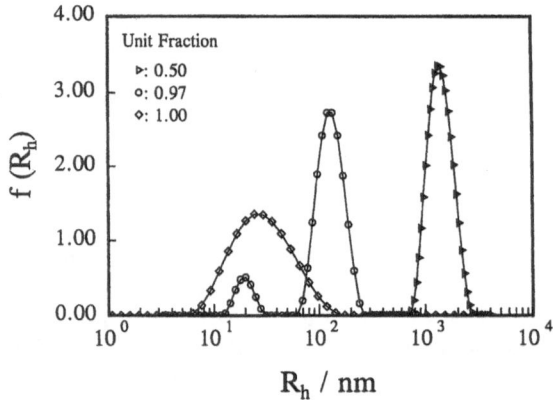

Fig. 23. Hydrodynamic radius distributions $f(R_h)$ for STVPh-9 and STVPh-9/PEMA with various blend compositions in toluene. The measuring conditions are the same as in Fig. 22 [147]

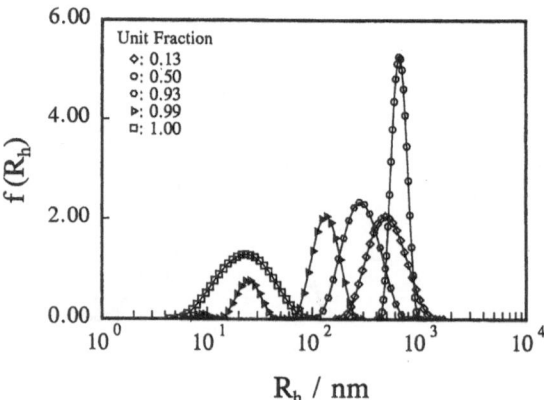

Fig. 24. Hydrodynamic radius distributions $f(R_h)$ for of STVPh-15 and STVPh-15/PEMA with various blend compositions in toluene. The measuring conditions are the same as in Fig. 22 [147]

of the free STVPh-9 to gradually disappear and the peak of the complex aggregates to move to a larger R_h. Finally, only a single peak appears at an R_h of ca.1000 nm for the STVPh-9/PEMA blend with a 1:1 base ratio, whose solution was unstable and began precipitating soon after mixing.

Formation of the complex aggregates becomes more apparent when the OH content is increased to 15 mol% (Fig. 24). Even when the added PEMA is as little as 1 wt%, it brings about a significant change in $f(R_h)$, with the appearance of a small peak at as large an R_h as 100 nm, which signifies complex aggregates. When the added PEMA reaches 7 wt%, $f(R_h)$ shows only a single peak, indicating that most STVPh-15 chains have aggregated with PEMA. When the fraction of STVPh approaches 0.5, the average R_h of the aggregates becomes even larger, while their $f(R_h)$ becomes narrower. It should be pointed out that complexation between STVPh and PEMA takes place and becomes detectable even at a concentration as low as $1.0\infty10^{-4}$ g/ml, which is 1/15th of the concentration used in viscosity measurements. A combination of LLS and NERT data thus allows us to conclude for STVPh/PEMA in toluene that intermolecular complexation occurs when the hydroxyl content in STVPh reaches about 9 mol%.

The results mentioned above illustrate the power of viscometry, NERT fluorospectroscopy and LLS for exploring intermolecular complexation, particularly its dependence on the contents of the interacting sites in the component polymers. The successful applications of these techniques can be seen in the studies using (1) PS(OH) as a proton-donating polymer and PMMA [111,143,144], PBMA [145], PCL [150,154], PEO [154], PES [154] as counter polymers, (2) STVPh as a proton-donating polymer and STVPy [146,153] and PEMA [147,152] as counter polymers, (3) PS(t-OH) and PS(s-OH) (Table 5) [158,159] as proton-donating polymers and poly(butyl methacrylate-co-4-vinyl pyridine) (BVPy) as a counter polymer [148], and (4) partially carboxylated polystyrene (CPS) [158,160] as a proton-donating polymer and BVPy as a counter polymer [149]. Table 7 summarizes the main results obtained on these polymer pairs in different solvents containing the minimum hydroxyl content required for complexation and allows us to draw the following conclusions:

1. All the systems studied undergo complexation strongly dependent on the content of the interacting sites. For a given system (proton-donating polymer, proton-accepting polymer and solvent), there exists a critical content of interacting sites below or above which unlike polymer chains either behave as independent coils or form complex aggregates. Each of the systems BVPy/PS(t-OH) and BVPy/CPS has two variables, namely, tertiary hydroxyl in PS(t-OH) or carboxyl in CPS as a proton donor and pyridyl (not butyl acrylate, Sect> 5.6) as a proton acceptor. The experimental results show that there is a critical value for proton acceptor content at a certain level of proton donor content and vice versa. However, we will not emphasize the critical values themselves, because they depend to some extent on experimental conditions such as the molecular weights of component polymers and temperature. Furthermore, the content intervals in the polymer samples used for the measurements may be too large to make an accurate estimation of the critical content. More important to recognize is that,

in dilute solutions of the blends with controllable hydrogen bonding, the transition from separate coils to complex aggregates takes place when the content of interacting sites and hence the density of intermolecular hydrogen bonds reaches a certain level. In some cases, the proton-donating content needed for this transition is only a few molar percent. This fact is not consistent with the traditional picture for the complexation of polymer pairs, according to which each segment of the polymer has its own interaction group and the polymer pairs form complexes of a ladder type [2].

Table 7. Polymer pairs forming complexes in solution

Proton acceptor	Proton donor	Minimum OH content	Solvent[a]	Method	Ref
PEMA	STVPh	9	toluene	LLS, NRET	[147,152]
PEMA	STVPh	22	PrNO$_2$		
PMMA	PS(OH)	8	toluene	Vis, NRET	[143,144]
PMMA	PS(OH)	no complexation	THF	Vis, NRET	
PBMA	PS(OH)	~10	toluene	Vis, NRET	[145]
PBMA	PS(OH)	no complexation	THF	Vis, NRET	
PCL	PS(OH)	8	toluene	Vis, LLS	[150,154]
		no complexation	THF		
STVPy-25	STVPh	6	CHCl$_3$	LLS	[146,153]
STVPy-50	STVPh	6	CHCl$_3$	Vis	
STVPy-25	STVPh	6	Butanone	LLS	
STVPy-25	STVPh	50	THF	Vis, LLS	
STVPy-50	STVPh	50	THF	Vis, NRET	
STVPy-75	STVPh	22	THF	Vis, NRET	
BVPy-45	PS(t-OH)	38	DCE	Vis, LLS	[148]
BVPy-51	PS(t-OH)	28	DCE	Vis, LLS	
BVPy-60	PS(t-OH)	21	DCE	Vis, LLS	
BVPy-68	PS(t-OH)	21	DCE	Vis, LLS	
BVPy-45	PS(t-OH)	no complexation	butanone	Vis	
BVPy-51	PS(t-OH)	no complexation	butanone	Vis	
BVPy-60	PS(t-OH)	29	butanone	Vis	
BVPy-68	PS(t-OH)	21	butanone	Vis	
BVPy-10	CPS	11	butanone	Vis	[149]
BVPy-20	CPS	7	butanone	Vis	
BVPy-29	CPS	7	butanone	Vis	
BVPy<29	CPS	no complexation	THF	Vis	
BVPy-29	CPS	11	THF	Vis	
BVPy-45	CPS	7	THF	Vis	

[a] THF: tetrahydrofuran, PrNO$_2$: nitropropane, DCE: dichloroethane

2. The concentrations used for viscometry, NRET fluorospectroscopy and LLS measurements were around $0.5-1\times10^{-2}$, 1×10^{-3} and 1×10^{-4} g/ml, respectively. Interestingly, although these concentrations differ considerably, the minimum contents of interacting sites for complexation estimated by these three techniques are similar. According to LLS studies [152,153], the concentration has an influence on the growth rate and final size of the complex aggregate but not on the critical content of interacting sites. This fact confirms that the contents of interacting sites in the complementary polymers are the determining factor for complexation.

3. As noted in Sects 2 and 3, one often identifies precipitates formed in mixing component polymers as a complex and the mixture cast from a transparent solution as a "blend". This convention should be accepted with reservation, since, in many cases, especially in LLS measurements, polymer complexes may exist in homogeneous solutions. What we measure in such cases is actually the behavior of a soluble complex.

4. The data shown in Table 7 indicate strong solvent effects on complexation. For example, PS(OH)/PMMA forms a complex in toluene but not in THF, while BVPy/CPS forms a complex in THF but the complexation needs a much higher content of interacting sites than in butanone. This kind of solvent effect as a basic feature of complexation is discussed in detail in Sects 5.5 and 6.

5.3
Time Dependence of Complexation [148,152]

In most cases of complex formation in polymer pairs in which each segment possesses an interaction site, complex precipitates are formed immediately after mixing, so that the process of complexation is too fast to be measured. However, in the case where the number of interacting sites is only slightly higher than the critical value and one of the component polymers is in substantial excess, it takes a few hours for the complex to settle down to a stable dispersion, so that the time evolution of complexation can be monitored. Figure 25 illustrates the evolution of the translational diffusion coefficient distribution G(D) for STVPh-9/PEMA in toluene. The initial broadening of the peak in G(D) reflects the formation of a small amount of STVPh-9/PEMA clusters. The clusters gradually gather to form larger ones. After ~70 min, two distinct peaks (1 and 2) appear in G(D). Peak 1 corresponding to the size of ~20 nm may be assigned to individual STVPh-9 chains, and peak 2 corresponding to the size of 150–300 nm to STVPh-9/PEMA complex aggregates. With increasing time, peak 2 shifts to the left and its area increases whereas peak 1 stays in the same position and its area decreases. This behavior shows gradual incorporation of individual STVPh-9 chains into STVPh-9/PEMA complexes. It should be noted that since the light-scattering intensity is proportional to the square of the scatter's mass, the large area of peak 2 actually reflects only a small number of STVPh-9/PEMA complexes.

For a blend solution with one component in excess, it is also possible to measure the apparent average hydrodynamic radius $<R_h>$ and the relative Rayleigh

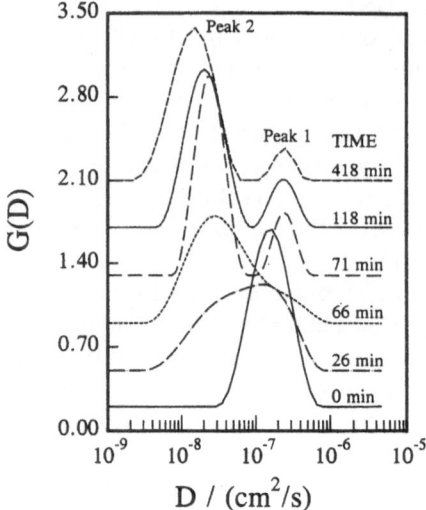

Fig. 25. Time evolution of the translational diffusion coefficient distribution G(D) for blend STVPh-9/PEMA after mixing, where the total polymer concentration is 1.0×10^{-4} g/ml, the weight ratio of STVPh-9/PEMA is 100:8, and the scattering angle is 15° [152]

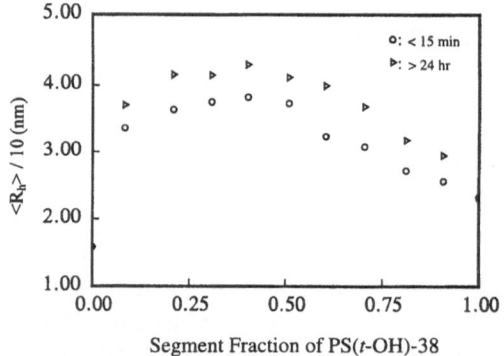

Fig. 26. Hydrodynamic radius $<R_h>$ for the blend PS(t-OH)-38/BVPy-60 in DCE as a function of segment fraction of PS(t-OH)-38 [148]

ratio $R_{vv,t}(q)/R_{vv,t=0}(q)$ as a function of time. The linear dependence of $[R_{vv,t=t}(q)/R_{vv,t=0}(q)]$ on time indicates the increase in the weight-average molar mass of the complexes at a constant rate and suggests that the complexation is diffusion controlled. It was found at a fixed STVPh-9 concentration of 1×10^{-4} g/ml that the larger the concentration of PEMA, the larger the rate constants of both $<R_h>$ and $R_{vv,t=t}(q)/R_{vv,t=0}(q)$ became.

The blend PS(t-OH)-38/BVPy-60 forms stable complexes in the inert solvent dichloroethane [148]. Figure 26 shows its $<R_h>$ measured as a function of the molar segment fraction of PS(t-OH)-38 at 15 min and 24 h after mixing. The plotted $<R_h>$ are all much larger than those of the component polymers, indicating the formation of polymer aggregates. The data also show a gradual growth of the complex aggregate with time, giving a size increase of about 15% in 24 h.

5.4
Composition of Complex Aggregates [152,153]

The composition of a complex formed in solution is conventionally determined by analysis of the complex precipitate or assumed to be the one at the minimum in the plot of viscosity vs. composition. For STVPh/PEMA in toluene, LLS gave the curves of $<R_h>$ against composition whose maxima appeared at the 1:1 base ratio of STVPh to PEMA, at which the corresponding viscosity curves showed a minima [147]. This finding means that the complexation attained the largest degree at this composition. Furthermore, since STVPh/PEMA in toluene may form a very stable dispersion when one component is in excess, LLS makes it possible to determine the composition and structure of the complex aggregates. Thus it was found from combined dynamic and static LLS measurements [152] that complexes whose $<R_h>$ was about 10 times larger than that of the individual STVPh-15 chain were formed when a very small amount (0.25–1%) of PEMA relative to STVPh-15 was added to STVPh solutions and that the average number of STVPh-15 in a complex particle was about 35 over this composition range. This appears to indicate that PEMA chains act as nuclei for complexation and that despite the fact that PEMA is only about 1 wt% relative to STVPh-15 was about 20 wt% if STVPh-15 was incorporated in the complex. When more PEMA was added, the size of complex aggregates grew more, probably as a result of further association of the particles.

Similar results were obtained for STVPh-50/STVPy-50 in THF [153]. For example, in the solutions containing only STVPh-50 of 0.1–1 wt% relative to STVPy-50, each complex aggregate contained one STVPh chain and 50–80 STVPy chains.

5.5
Solvent Effect on Complexation [111,143–150,152–154, 159]

The data in Table 7 for the blends with controllable hydrogen bonding clearly show solvent effects. The η_{sp}/c vs. composition curve for PS(OH)/PMMA [143] in toluene in Fig. 20 shows two kinds of behavior attributable to separate coils and complex aggregates, depending on the hydroxyl content in PS(OH). However, this blend in THF is transparent and shows η_{sp}/c which varies with the composition nearly obeying the weight-average law, regardless of the OH content in PS(OH). Thus, no complex can be formed in THF [143]. These results are consistent with the ones from NRET fluorospectroscopy shown in Fig. 21, where the values of I_c/I_a for carbazole-labeled PS(OH) and anthracene labeled PMMA in toluene display two levels depending on the hydroxyl content in PS(OH). The lower values of I_c/I_a corresponding to the complexes in these solutions increase to the same value when a small amount (<1%) of THF is added, which indicates that the complex formed in toluene dissociates with the addition of THF, owing to the stronger hydrogen-bond-forming ability of THF than that of PMMA with PS(OH).

THF is also less favorable for complexation between STVPh and PEMA, where the phenol groups serve as proton donors [147]. Figure 27, which illustrates the variation of I_c/I_a for STVPh/PEMA in toluene with the amount of THF added, shows three different types of behavior. (1) At very low OH contents (1–3mol%), I_c/I_a stays at a high level almost independent of the THF concentration, which indicates that the component polymer chains remain separated and the situation does not change with the addition of THF. (2) When the OH content in STVPh is increased to 6 mol%, I_c/I_a increases a little with the initial addition of THF, which may be due to the dissociation of interpolymer hydrogen bonding. (3) When the OH content is further increased to 9 mol% or more, a marked increase in I_c/I_a with the addition of THF occurs and simultaneously the originally turbid mixture becomes clear, which evidences a decrease in the interpenetration of the unlike chains due to disruption of hydrogen bonding. Finally, as more THF is added, all the blends show nearly the same high I_c/I_a values, and the complexes between STVPh and PEMA are actually destroyed. Figure 27 also clearly shows that the higher the OH content the more THF is required to destroy the complexes. Jiang et al. have shown that THF completely depresses complex formation in PS(OH)/poly (alkyl methacrylate) [143,145], PS(OH)/PCL [154], STVPh/poly(alkyl acrylates) [147], PS(t-OH)/BVPy [148] and PS(s-OH)/BVPy [159]. However, STVPh/STVPy [146] and CPS/BVPy [149] form complexes in THF, although much higher contents of interacting sites are needed than in inert solvents.

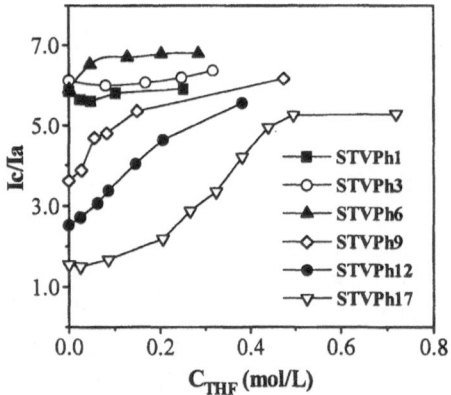

Fig. 27. Ic/Ia of STVPh/PEMA blend (50:50, w/w) in toluene as a function of the amount of added THF, for different OH contents in STVPh. The total polymer concentration is 1.0×10^{-4} g/ml. Concentrations of carbazole in STVPh and that of anthrancene in PMMA are 0.52×10^{-5} and 1.31×10^{-5} M, respectively [147]

Tsuchida and Abe [2] presented a simple scheme which assumes that the occurrence of complexation depends on the competition of the interaction forces E_{ad} and E_{sp}, where E_{ad} denotes the force between a pair of proton-accepting and proton-donating polymers and E_{sp} the forces between solvent and polymer. The net driving force for polymer-polymer complexation is thus written as:

$$E_{complex} = E_{ad} - E_{sp} \tag{11}$$

Complexation occurs when the favorable force, E_{ad}, overcomes the unfavorable force E_{sp}. This idea may be crude, yet it is a useful guideline in understanding the basic factors controlling complexation. First, in any given system, there should exist a certain critical content of interacting sites, which defines the change from separated coils to complex aggregates. This is expected because increasing interaction sites in proton-donating polymers should increase E_{ad}. Second, pronounced solvent effect reflects the competition between the solvent and one of the component polymers for the formation of hydrogen bonds.

In systems such as PS(OH)/PMMA, PS(OH)/PCL and STVPh/PEMA, the proton-accepting ability of THF to form hydrogen bonds with the hydroxyl groups is generally stronger than that of the carbonyl in proton-donating polymers, so that no polymer-polymer complex may be formed in THF. However, in STVPh/STVPy and CPS/STVPy, pyridyl, a strong proton-accepting group, favors the hydrogen bonding with both phenol of STVPh and carboxyl of CPS over that between THF and the proton donor in the polymers. Actually, the stable complex between STVPh and STVPy in THF can be destroyed by use of even stronger proton-accepting solvents such as DMF and water, as has been evidenced by LLS and NRET [146,153].

Figure 28 shows the water content dependence of the relative scattering intensity I/I_0 and the average hydrodynamic radius $<R_h>$ for the STVPh-50/STVPy-

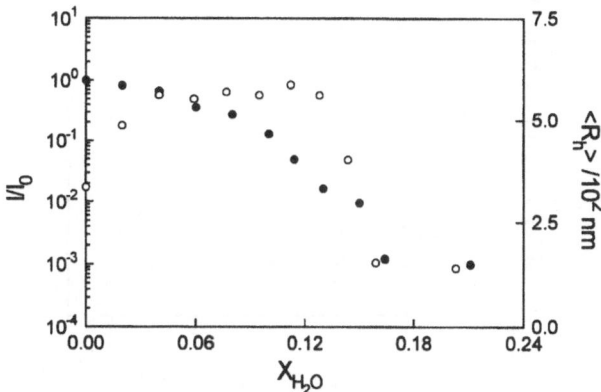

Fig. 28. Water content dependence of the relative intensity (I/I_0) (●) and average hydrodynamic radius $<R_h>$ (○) for equimolar STVPh-50/STVPy-50. The total polymer concentration is 1.0×10^{-4} g/ml [153]

50 complexes in which the molar ratio of the monomer units of STVPh-50 to STVPy-50 is 1:1 in THF, with X_{H2O} denoting the molar fraction of water added and I_0 the scattering intensity before addition of water. The substantial decrease in I/I_0 indicates the dissociation of complexes by the addition of water; note that water has a stronger hydrogen-bonding ability than THF, and hence it tends to destroy hydrogen bonding between STVPh and STVPy. Differing from I/I_0, $<R_h>$ initially increases with X_{H2O}. This behavior indicates the swelling of the complex prior to dissociation. The two curves show breaks at X_{H2O} of about 0.12, where decomplexation starts, and level off at X_{H2O} about 0.16, where no polymer-polymer complex exists any more.

It has been found that CPS is not miscible with poly(butyl methacrylate). Hence, the effective interacting sites in CPS/BVPy must be carboxyl in CPS and pyridyl in BVPy [149]. A series of CPS/BVPy blends was prepared and studied in butanone and THF by viscometry. CPS is not soluble in a typical inert solvent like toluene, but is soluble in butanone, which may have a weak effect on hydrogen bonding. As mentioned in Sect. 5.1, the states of non-complexes and complexes can be easily differentiated by viscosity profiles. Figure 29 shows a 'complexation map' for CPS/BVPy in butanone and THF, made on the basis of viscosity data. It displays the boundaries between non-complex and complex regions for both solvents. First, the carboxyl content in CPS and the pyridyl content in BVPy affect complexation and that complexation takes place only when the hydrogen-bonding density exceeds a certain level. Second, we see the solvent effect that when compared with the blends in butanone, those in THF possess smaller 'complex' areas. Thus, CPS-11 can complex with BVPy having 10 mol% pyridyl content or more in butanone but 30 mol% in THF.

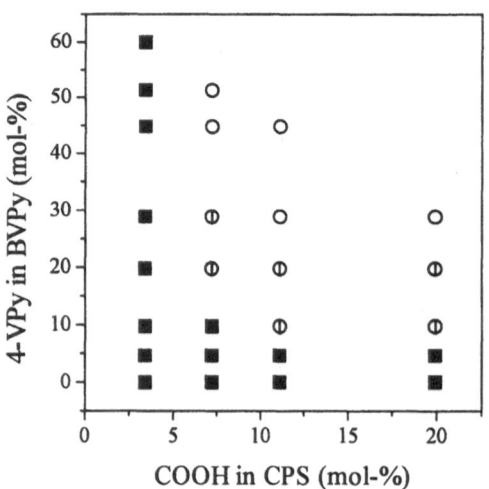

Fig. 29. Complexation map for CPS/BVPy blends in 2-butanone and THF. No complexation in both solvents (■), complexation in 2-butanone but not in THF (◐), complexation in both solvents (○) [149]

5.6
Self-association vs. Inter-association [148,149,159]

Among some of the proton-donating polymers listed in Table 5, i.e. CPS, PS(OH), PS(t-OH) and PS(s-OH), CPS differs from the others by its strong ability of self-hydrogen bonding, as can be inferred from the order of magnitude of the PKa and K_2 of their monomers in bulk at 25 °C [161], i.e.:

pKa=15(HPS-HES)>10(4-VPh)>7.4(HFS)>5(4-VBA)

K_2=173,000(4-VBA)>21(4-VPh)>2.32(HFS)

Here, HPS stands for p-(2-hydroxyisopropyl)styrene, HES for p-(1-hydroxyethyl)styrene, 4-VPh for 4-vinyl phenol, HFS for p-(hexafluoro-2-hydroxyisopropyl)styrene, and 4-VBA for 4-vinylbenzoic acid. Among these, 4-VBA has the lowest pKa and the highest K_2, which implies that the carboxyl group has the strongest ability to form a dimer and to liberate its proton simultaneously. Thus, we can expect that CPS forms ring-like dimers and shows miscibility and complexation features intrinsically different from PS(OH), PS(t-OH), PS(s-OH) and STVPh.

Viscosity measurements on a series of CPSs having the same polymerization degree but different carboxyl contents have provided clear evidence for self-association in solutions. Figure 30 shows the reduced viscosities in butanone and THF (2 mg/ml) at 30.0 °C as a function of carboxyl content. When the carboxyl content is below 3.4 mol%, the reduced viscosity decreases as carboxyl groups are introduced into the CPS. This behavior reflects the constriction of polymer

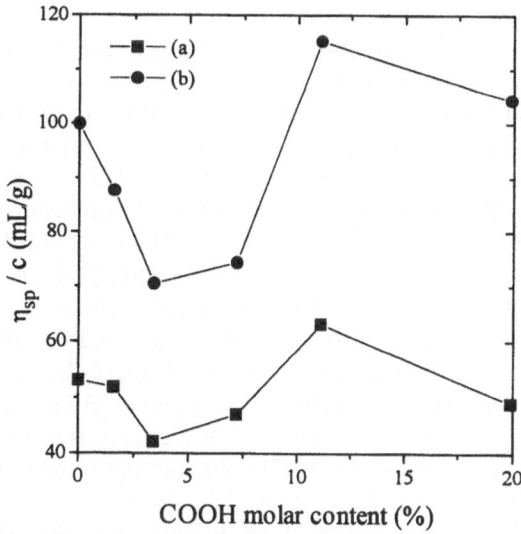

Fig. 30. Reduced viscosity of CPS in *a* 2-butanone and *b* THF as a function of carbonyl molar content in CPS [149]

coils due to intramolecular association of the carboxylic acids. Then, this reduced viscosity turns to an increase with the increase in the carboxyl content, as
a result of the suppression of intramolecular association by competing intermolecular association. After reaching a maximum at a carboxyl content of about
11 mol%, the reduced viscosity falls owing to the decrease in solubility, the CPS
with the higher carboxyl content suffers.

The effect of strong self-association of a proton-donating polymer on its complexation with proton-accepting polymers can be interpreted in terms of the
competition of E_{ad}, E_{sp} and the self-association force E_{dd} of the proton-donating
polymer. Thus, the net driving force for interpolymer complexation is given by:

$$E_{complex} = E_{ad} - E_{sp} - E_{dd} \tag{12}$$

This expression indicates that for a polymer having strong self-association
power a greater polymer-polymer interaction is needed for complexation. In
fact, although CPS has a strong proton-donating ability as its low PKa suggests,
it is not miscible with PBMA, and much stronger proton-accepting pyridyl is
needed to get it miscible and complexed. Table 7 shows the critical contents of
carboxyl in CPS for forming complexes with BVPy having different pyridyl contents. When E_{dd} is comparable to E_{sp} and E_{ad}, intramolecular hydrogen bonding
may still exist in the complex aggregates, so that part of the hydroxyl groups is
"wasted" in intramolecular hydrogen bonding, and loops may be formed to
render the complex a loose structure. This kind of complex precipitate often has
the base ratio of proton-donating polymer to proton-accepting polymer larger
than 1:1. However, if a solvent with a higher E_{sp} is used, the intramolecular association may be destroyed, and the complex may possess a composition close
or equal to 1:1 [149,159].

There have been conflicting opinions about the interaction between carboxyl
and pyridyl groups in polymer blends. Some authors considered it an acid-base
or proton-transfer interaction [162,163], while others considered it the hydrogen-bond interaction [164–167]. The FTIR spectrum of the CPS-11/BVPy-29
blend shows a characteristic absorption band at 1557 cm^{-1} [149], assigned to the
C=C stretch in the pyridine ring [55], but no absorption band around 1635 cm^{-1}
[167,168] characteristic of the pyridinium ion. Zhu et al. [149] took this fact as
indicating that the interaction between carboxyl and pyridyl groups is mainly of
hydrogen-bonding nature, not of proton-transfer interaction.

PS(t-OH) having tertiary hydroxyl groups tends to suffer both weak self- and
inter-association. The presence of electron-repulsing -CH$_3$ lowers its proton-donating ability as suggested by the high PKa value of HPS, and the presence of
bulky (CH$_3$)$_2$(OH)C- groups gives rise to steric effects. It was observed that the
reduced viscosity for a series of PS(t-OH), having the same polymerization degree but different functionality, decreased in the almost inert solvent dichloroethane (DCE), but remained unchanged in butanone when the hydroxyl content was increased. The latter behavior may be ascribed to the disappearance of
weak self-association [148,159].

Being a weak proton-donating polymer, PS(t-OH) does not form either a miscible or complex blend with PBMA, but can complex with BVPy of high pyridyl contents. The large difference in proton-donating ability between the tertiary hydroxyl in PS(t-OH) and the carboxyl in CPS can be clearly seen from their immiscibility-miscibility-complexation maps, which will be discussed in Sect. 6.2.

6
Complexation of Blends with Controllable Hydrogen Bonding in the Solid State

In phase relationship, ordinary miscible blends and complex blends are indistinguishable because both appear as a single phase. The routine techniques characterizing phase structure such as DSC, TEM and NMR relaxation time measurements are generally incapable of elucidating chain arrangements at the molecular level. It is generally believed that the chains of different components are randomly mixed in ordinary miscible blends, while unlike segments are paired together in polymer complexes. Although this difference is conceptually clear, its experimental confirmation has not been easy to achieve. In fact, even for such extensively studied "classical" complex systems as PAA/PEO and PAA/PVPo, we still know very little about their structural characteristics in the solid state. In this section, we discuss some of the results obtained for solid blends with controllable interpolymer hydrogen bonding by using fluorescence techniques and thermographic analysis. They give us some insight into the difference at the molecular level between ordinary miscible blends and complexes.

6.1
Complexation in the Solid State

6.1.1
Evidence from NRET Measurements

As discussed in Sect. 4.3, NRET fluorospectroscopy has been successfully used for monitoring the transition from immiscibility to miscibility in toluene-cast PS(OH)/PMMA film that accompanied the increase in hydrogen-bonding density. This study has been extended to blends containing hydroxyls up to 17 mol% in PS(OH) and has led to new findings [92,108]. Two series of blends, PS(OH)/PMMA and PS(OH)/PBMA containing anthracene and carbozole fluorescent labels, were studied, with the results shown in Fig. 31. Here we see that the two series of blends give nearly parallel variations of Ic/Ia, with hydrogen-bond density. The most remarkable feature is the appearance of two transitions and two plateaus. An additional measurement was made on a miscible reference blend consisting of anthracene-labeled PMMA and carbazol-labeled PMMA, at the same concentration of fluorescence probes as that in the two series of blends, and it gave 0.33 for Ic/Ia. This value is comparable with the first plateau values [0.33 for PS(OH)/PBMA and 0.25 for PS(OH)/PMMA] in Fig. 31 and confirms

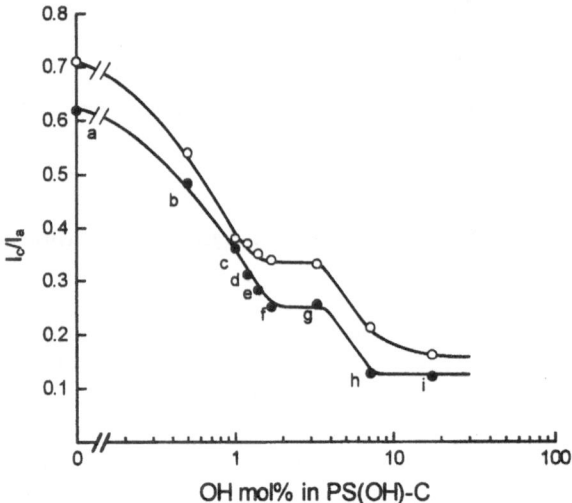

Fig. 31. Variation of Ic/Ia for blends PS(OH)/PMMA (50:50, w/w) (●) and PS(OH)/PBMA (○) as a function of the hydroxyl content in PS(OH). PS(OH) is labeled with carbazole (donor), and PMMA and PBMA are labeled with anthrancene (acceptor). Concentrations of carbazole and anthrancene are 5.17×10^{-3} M [108]

the miscibility of these blends at the first plateau. Thus, in both cases, the decrease in Ic/Ia with further increase in the hydroxyl content from 3 to about 7 mol% may be taken as indicating that the segment contact between unlike components becomes more favored than that between like components. Considering the large body of experimental results for the solutions of PS(OH) and poly(alkyl methacrylate) mentioned in Sect. 5 [143–145], we can reasonably attribute the unusually large efficiency of energy transfer to the complexation between the proton-donating and proton-accepting polymers.

It is interesting to note that, in most cases studied, the range of hydroxyl content in PS(OH) which allows the miscibility-to-complex transition to take place in solvent-cast films is almost the same as that allowing separate coils to transform to complex aggregates in solutions [143]. This agreement suggests that complexes formed in dilute solutions remain undestroyed during the process of solvent evaporation.

6.1.2
A New NRET Approach to Monitor Complexation

In Fig. 31 we see that the decrease in Ic/Ia over the probable transition from miscibility (0.3) to complexation (0.15) is much smaller than that over the transition from immiscibility (0.6) to miscibility (0.3). Therefore, the NRET method has been modified in order to differentiate more clearly molecular arrangements be-

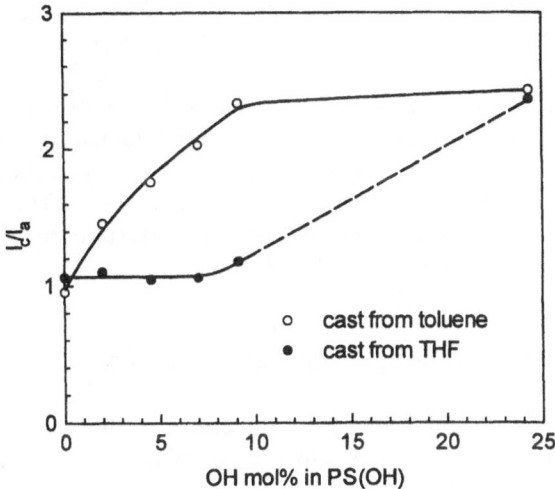

Fig. 32. Ic/Ia for blends PS(OH)/PMMA-a/PMMA-c cast from toluene (○) and from THF (●) as a function of the hydroxyl content in PS(OH). PMMA-a and PMMA-c refer to PMMA labeled with anthrancene and carbazole, respectively [111]

tween an ordinary miscible blend and a polymer complex by partially labeling one component with a donor and an acceptor, instead of attaching a donor to one component and an acceptor to the other [111,169]. Among the blends made of a donor-labeled polymer A, an acceptor-labeled polymer A', and a chromophore-free polymer B, the immiscible one will have the largest efficiency of energy transfer, because chains A and A' are concentrated and randomly mixed in their own phase. The efficiency is expected to decrease when the blend becomes miscible since the labels are "diluted" with polymer B. A more pronounced decrease of the efficiency should occur upon complexation because A and A' then turn to bind with B in such a way that the donor and acceptor are to some extent "isolated" by B. This efficiency change will become even more marked when A is the minor component.

Figure 32 presents the relevant experimental data concerning these predictions, obtained with the blends of anthracene-labeled PMMA, carbazole-labeled PMMA and PS(OH) cast from toluene. The initial value of Ic/Ia (1.0) is for the immiscible blend with no hydrogen bonding. Ic/Ia increases to 1.4 as the OH content in PS(OH) increases to 2.0 mol%, at which the blend is miscible as evidenced by DSC and TEM. It monotonically increases with further increase in the OH content and suddenly levels off to a value of 2.4 at an OH content of 9 mol%. This change in Ic/Ia accompanying the increase in hydrogen-bonding density above the value corresponding to the immiscibility-miscibility transition can be taken as evidence for the difference in chain arrangements between "ordinary" miscible blends and polymer complexes.

Figure 32 also shows that the OH dependence of Ic/Ia is distinctly different for PMMA/PS(OH) blends cast from toluene and THF. Ic/Ia for THF-cast films almost stay at the value (1.05) of the immiscible blend until the hydroxyl content reaches 7 mol%, and increases a little when the hydroxyl content goes up 9 mol%, which indicates inception of miscibility. Then, a considerable increase of Ic/Ia occurs to 2.4, a value comparable to that for toluene-cast complex blends, when the hydroxyl content increases to as high as 24 mol%. Thus, a much higher hydroxyl content (24 mol%) is needed to allow interpolymer complexation to take place in THF than in toluene (9 mol%).

6.2
A Map of Immiscibility-Miscibility-Complexation Transitions

The important conclusion drawn from the above studies on PS(OH)/PMMA in solution and bulk is that complexes formed in dilute solutions can be preserved during the process of film casting. In particular, when we use an inert solvent whose E_{sp} is close to zero, the critical hydroxyl contents in proton-donating polymers for complexation estimated by viscosity or LLS are comparable to that for the miscibility-to-complex transition in bulk, which can be easily detected by DSC or TEM. Therefore, by combining the results from both solution and bulk, it should be possible to construct a map for a given blend system visualizing how the immiscibility, miscibility and complexation of the blend depend on the content of interacting sites.

Here, we present such a map for the blends composed of PS(t-OH) and BVPy. In this system, two variables, i.e. the contents of tertiary hydroxyl in PS(t-OH) and pyridyl in BVPy, control the interaction of the polymer components [148]. Blends of 70:30, 50:50 and 30:70 in weight composition, each covering a range of these variables, were prepared, their phase behavior was examined by DSC, and the transition from separate coils to complex aggregates in the inert solvent dichloroethane was detected by viscosity and LLS. Figure 33 shows the resulting immiscibility-miscibility-complexation transition map. We see that the blend is immiscible only when one or both components have very low contents of interacting sites and that there is a large area of miscibility. In addition, complexation can take place only for blends whose components have relatively high contents of interacting sites. Furthermore, in the PS(t-OH)-38 series, the minimum pyridyl content in BVPy for complexation is as high as 45 mol%. If the hydroxyl content in PS(t-OH) is less than 8 mol%, the blend forms no complex, no matter how high the pyridyl content in BVPy is.

Figure 34 shows the corresponding map for the blend CPS/BVPy. Since CPS does not dissolve in any inert solvents, butanone was used for solution measurements as well as for film casting. The figure displays three regions representing the immiscible, ordinary miscible, and complex blends. When the COOH contents in both CPS and BVPy are low, or either of them is low, the blend is immiscible, but once the hydrogen bond density exceeds a certain critical level, it turns to be miscible.

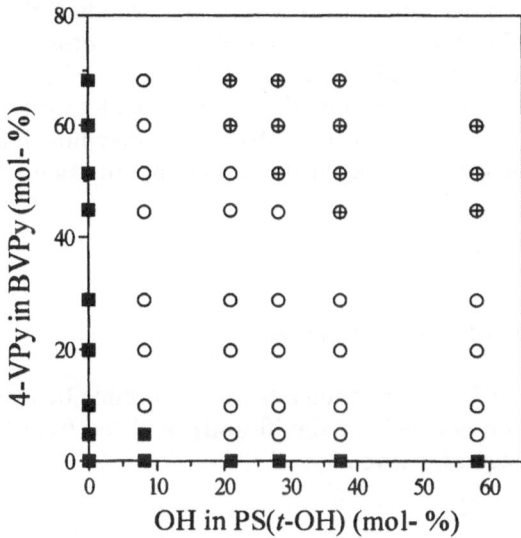

Fig. 33. Map of immiscibility-miscibility-complexation transitions for PS(t-OH)/BVPy (■) immiscible, (○) miscible and (⊕) complexed [148]

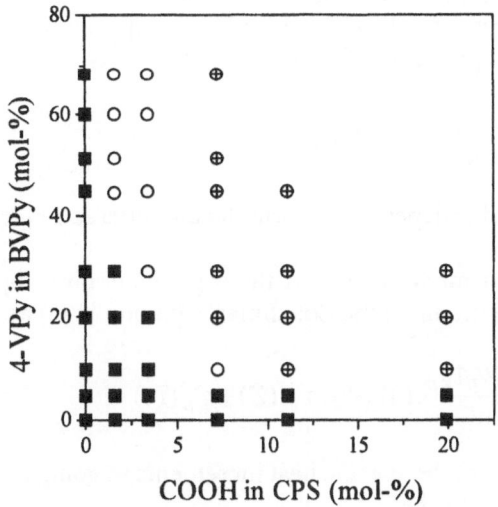

Fig. 34. Map of immiscibility-miscibility-complexation transitions for CPS/BVPy (■) immiscible, (○) miscible and (⊕) complexed [149]

The most substantial difference between the blends CPS/BVPy and PS(t-OH)/BVPy lies in the ability to form complexes. The proton donor in PS(t-OH) is much weaker than that in CPS, so that much higher interaction site concentrations are required for complexation. For example, BVPy with pyridyl less than 40 mol% does not complex with PS(t-OH), no matter how high the hydroxyl content is. However, BVPy with a pyridyl content of only 10 mol% may complex with CPS containing carboxyl more than 11 mol%.

6.3
Glass-Transition Behavior of Complex Blends

It is known that the T_g of binary blends made miscible by hydrogen bonding shows a composition dependence significantly deviated from the weight-average law or the Fox equation given by:

$$1/T_g = W_A/Tg_A + W_B/Tg_B \tag{13}$$

where W_A and W_B are the mass fractions and Tg_A and Tg_B are the glass-transition temperatures of components A and B, respectively.

Generally, hydrogen bonding raises T_g since it restricts the motion of polymer segments. However, this does not mean that the increase in T_g always appears as a positive deviation from the weight-average law. For example, the T_g-composition curves of miscible blends consisting of PS and poly(vinyl methyl ether) (PVME) display negative deviations from this law. Although the introduction of hydrogen bonding into PS somewhat increases the T_g of the blend, the Tg-composition relation for STHFS/PVME still appears below the weight-average law [85]. Generally, the T_g-composition curves of hydrogen-bonded blends are either convex or sigmoidal. Kwei [61,170] have proposed the following equation for composition dependence of T_g for miscible blends:

$$T_g = \frac{W_A Tg_A + kW_B Tg_B}{W_B + kW_B} + W_A W_B q \tag{14}$$

where q is assumed to depend on intermolecular interaction and k is a fitting parameter.

By developing a model based on the arguments given by Goldstein [171], Painter et al. [172] modified the Couchman equation [173] as follows:

$$T_g = \frac{W_A Tg_A + W_B Tg_B}{W_A + kW_B} + W_A W_B (q'_m (X) + q'_B (T)) \tag{15}$$

where k is the ratio of the specific heat increments in going from the glass to the liquid state for A and B, i.e.

$$k = (\Delta C_p)_B / (\Delta C_p)_A \tag{16}$$

with A and B denoting proton-accepting and proton-donating polymer, respectively, and the q' terms can be calculated from the spectroscopically determined equilibrium constants. Note that Eq. (15) has no adjustable parameter. Painter et al. [83] found that it gives a composition dependence of T_g consistent with experiments for PVPh/PVPy and PHMP/PMMA, which show convex and sigmoid relations, respectively.

Usually, the T_g of a complex blend is higher than that of the corresponding miscible blend. Zhu et al. [148] compared the T_g-composition relationships of ordinary miscible blends and complex blends. Typical results are shown in Fig. 35 for a series of blends consisting of PS(t-OH)-38 and BVPy containing pyridyl from 9.8 to 68 mol%. These blends are either miscible or complexed, each having one T_g. The T_g vs. composition relationships shown are classified into two types. When the hydrogen-bonding interaction is relatively weak as in BVPy containing pyridyl below 45 mol%, the relationship is sigmoidal, giving T_g below and above the weight-average law at lower and higher PS(t-OH) contents, respectively. On the other hand, when the pyridyl content is high, as in BVPy-60, the relationship is convex, showing a positive deviation from the weight-average law over the whole composition range. However, even though these two types of relationships are roughly characteristic of ordinary miscible blends and complexes, this distinction alone cannot be taken as an criterion for complexation, as can be seen from Fig. 36, which concerns CPS-20/BVPy. When the 4-VPy content in BVPy is below or equal to 20 mol%, the T_g vs. composition relationship is sigmoidal. In fact, viscometry data indicate that CPS-20 may complex with all BVPy having pyridyl contents larger than 10 mol%. This sigmoidal behavior is

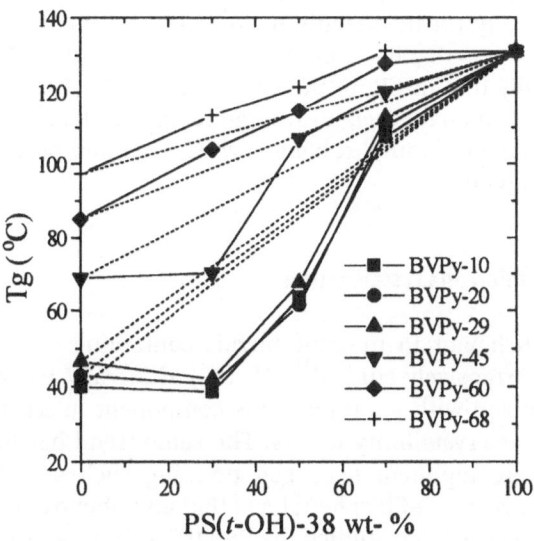

Fig. 35. T_g of PS(t-OH)-38/BVPy-x as a function of the weight composition of the blend. *Dashed lines* represent the weight-average law values [148]

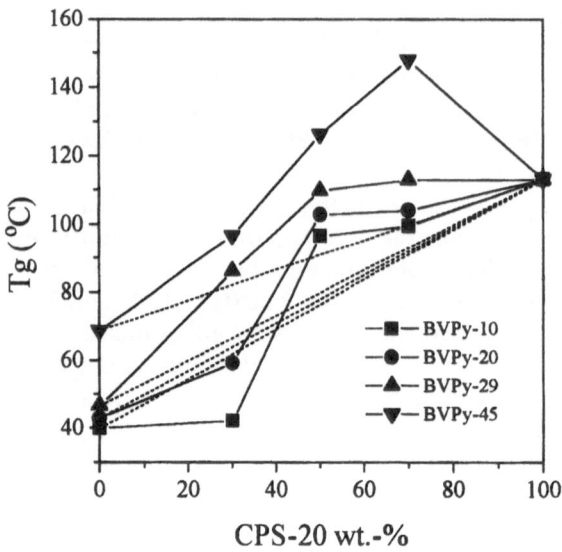

Fig. 36. T_g of CPS-20/BVPy-x cast from 2-butanone as a function of the weight composition of the blend. *Dashed lines* represent the weight-average law values [149]

understandable because when the CPS content is less than 40%, the total number of the carboxyls may not match the need for allowing all BVPy chains to take part in complexation owing to the self-association of carboxyl, though the number of carboxyls in each CPS chain is comparable to that of VPy in each BVPy chain. Actually, the blend with the low CPS content is a mixture of a complex and free BVPy, so that its T_g appears below the weight average.

It is noteworthy that the T_g of the CPS-20/BVPy-45 (70:30, w/w) complex is about 50 °C higher than the weight-average value. This fact allows us to expect a possibility of producing complex blends which are more heat resistant than ordinary miscible blends.

6.4
Effect of Complexation on Crystallization

Crystallization behavior in miscible blends containing crystallizable components has been extensively studied [174–180]. Generally, when a crystallizable component is mixed with an amorphous component its melting temperature goes down and its crystallinity lowers. The same trend has been reported for blends with intercomponent hydrogen bonding such as PCL/STVPh [181], PCL/poly(hydroxyl ether of bisphenol A) [182] and phenoxy resin/PEO [183].

Recently, Zhou et al. [150] studied the PS(OH)/PCL system, with special interest in the effect of complexation on the crystallization of PCL. A miscibility search by DSC showed that PS(OH)-3/PCL was miscible over a limited composi-

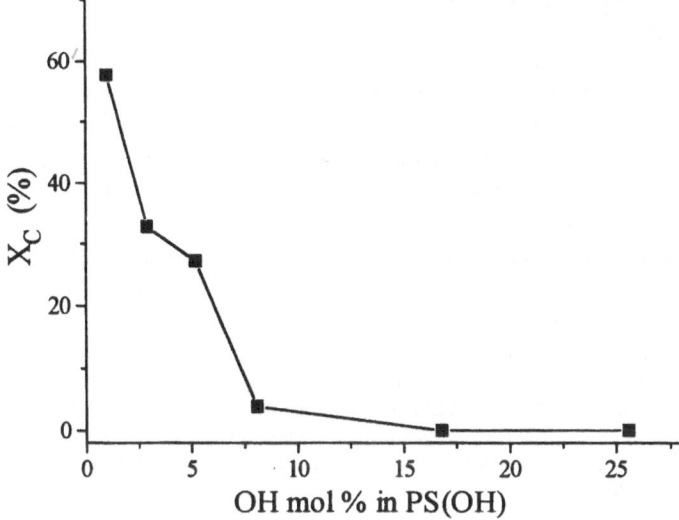

Fig. 37. Crystallinity Xc of PCL as a function of the hydroxyl content in PS(OH)-x for the annealed blend PS(OH)-x/PCL (60:40, w/w) [150]

tion range but PS(OH)-5/PCL was miscible over the whole range. Furthermore, viscosity and LLS measurements revealed that the minimum hydroxyl content in PS(OH) for complexation with PCL in the inert solvent toluene was about 8 mol%.

Figure 37 shows the crystallinity Xc of PCL plotted against the hydroxyl content in PS(OH) for the blend PS(OH)/PCL 60:40. The composition 60:40 is close to the one at which the reduced viscosity exhibits a minimum, so that the blend is expected to have a maximum degree of complexation. Xc was calculated from:

$$X_C = \frac{\Delta H}{\Delta H^o \phi} \times 100 \qquad (17)$$

where ΔH^o, the melting heat of 100% crystalline PCL, is 136.08 J/g [184] and ϕ is the weight fraction of PCL in the blend.

As shown in Fig. 37, Xc markedly decreases with the increase in the hydroxyl content in PS(OH), although not regularly. For PS(OH)-1/PCL, a typical immiscible blend, Xc is 0.58, which comes close to the Xc of pure PCL (0.73) prepared under the same annealing conditions. A substantial decrease in Xc appears for PS(OH)-3/PCL and PS(OH)-5/PCL, both judged to be miscible by T_g analysis. Crystallinity depressions similar to Fig. 37 have been reported on miscible amorphous/crystalline blends [182,183], although not quantitatively. The most interesting fact revealed in Fig. 37 is that crystallization of PCL is almost completely depressed in the blends with PS(OH) having hydroxyl contents of 8, 17 and 26 mol%. These blends are all exactly those found by viscometry to form complex aggregates in solution. A similar fact has been found from the study on another blend, STVPh/PCL, in which the hydroxyl content was varied over a broader range from 1 to 50 mol% [154]. All these findings seem to explain why the crystallizable components in complex blends undergo unusually large depression of Xc. In connection with this it is worth emphasizing that the main characteristics of a complex blend come from the pairing of unlike segments. For PCL segments to enter the crystalline lattice, they must first get rid of the pairing, and this needs cooperative dissociation of hydrogen bonds. At the crystalline temperature of PCL, it is very difficult for this process to occur, actually making crystallization impossible.

Acknowledgment. The research work of Jiang's group mentioned in this review has been supported by the National Natural Science Foundation of China (No. 28970183, 29374156, 29574154) and the National Basic Research Project-Macromolecular Condensed State. The light-scattering experiments were carried out in Prof. Chi Wu's laboratory at the Chinese University of Hong Kong.

7
References

1. Tsuchida E (1991) Macromolecular complexes: dynamic interactions and electronic processes. VCH, Weinheim
2. Tsuchida E, Abe K (1982) Adv Polym Sci 45:1
3. Bekturov E, Bimendina L (1981) Adv Polym Sci 41:99
4. Dubin P, Bock J, Davis R, Schulz D, Thies C (1994) Macromolecular complexes in chemistry and biology. Springer, Berlin Heidelberg New York
5. Hara M (1993) Polyelectrolytes: science and technology. Marcel Dekker, New York
6. Challa G (1990) In: Lemstra P (eds) Integration of fundamental polymer science and technology, vol 5. Elsevier Applied Sci, London, p 85
7. Rodriguez-Parada J, Percec V (1986) Macromolecules 19:55
8. Bekturov E, Bimendina L (1997) J Macromol Sci–Rev Macromol Chem Phys C37:501
9. Osada Y (1979) J Polym Sci Polym Chem Ed 17:3485
10. Tsuchida E, Osaka Y, Ohno H (1980) J Macromol Sci Phys B 17:683
11. Pèrez-Gramatges A, Argüelles-Monal W, Peniche-Covas C (1996) Polym Bull 37:127
12. Chatterjee S, Yadav J, Sethi K (1985) Angew Makromol Chem 130:55
13. Chen H, Morawetz H (1982) Macromolecules 15:1445
14. Chen H, Morawetz H (1983) Eur Polym J 19:923
15. Bednár B, Morawetz H, Shafer A (1984) Macromolecules 17:1634
16. Antipina A, Baranovskii V, Papisov M, Kabanov V (1972) Polym Sci USSR 14:1047
17. Ikawa T, Abe K, Honda K, Tsichida E (1975) J Polym Sci Polym Chem Ed 13:1505
18. Iliopoulos I, Audebert R (1985) Polym Bull 13:171
19. Bednár B, Li Z, Huang Y, Chang L, Morawetz H (1985) Macromolecules 18:1829
20. Kim B, Jeon S, Ree T (1986) Bull Korean Chem Soc 7:238
21. Iliopoulos I, Audebert R (1988) Eur Polym J 24:171
22. Iliopoulos I, Halary JL, Audebert R (1988) J Polym Sci Polym Chem Ed 26:275
23. Oyama H, Tang W, Frank C (1987) Macromolecules 20:474
24. Jeon S, Ree T (1988) J Polym Sci Part A Polym Chem 26:1419
25. Eagland D, Crowther N, Butler C (1994) Eur Polym J 30:767
26. Morawetz H (1965) Macromolecules in solution. Wiley, New York, p 363

27. Anufrieva E, Birshtein T, Nekrasova T, Ptitsyn O, Sheveleva T (1968) J Polym Sci 16:3519
28. Chu D, Thomas J (1984) Macromolecules 17:2412
29. Oyama H, Tang W, Frank C (1987) Macromolecules 20:1839
30. Frank C, Hemker D, Oyama H (1991) In: Shalaby S, McCormick C, Butler G (eds) Water soluble polymers. ACS Symposia Series 467, p 303
31. Baranovsky V, Shemkov S, Rashkov I, Borisov G (1992) Eur Polym J 28:475
32. Baranovsky V, Shemkov S, Rashkov I, Borisov G (1991) Eur Polym J 27:643
33. Petrova Ts, Rashkov I, Baranovsky V, Borisov G (1991) Eur Polym J 27:189
34. Baranovsky V, Petrova Ts, Rashkov I (1991) Eur Polym J 27:1045
35. Maltesh C, Somasundaran P, Pradip, Kulkarni R, Gundiah (1991) Macromolecules 24:5775
36. Sivadasan K, Somasundaran P (1991) J Polym Chem Polym Chem Ed 29:911
37. Sivadasan K, Somasundaran P, Turro N (1991) Colloid Polym Sci 269:131
38. Yasuda K, Okajima K, Kamide K, (1988) Polym J 20:1101
39. Heyward J, Ghiggino K (1989) Macromolecules 22:1159
40. Oyama H, Hemker D, Frank C (1989) Macromolecules 22:1255
41. Anufrieva E, Gotlib Yu, Krakoviak M, Skorokhodov S (1972) Polym Sci 14:1604
42. Bokias G, Staikos G, Iliopoulos I, Audebert R (1994) Macromolecules 27:427
43. Krupers M, Van Der Gaag F, Feijen J (1996) Eur Polym J 32:785
44. Iliopoulos I, Audebert R (1988) J Polym Sci Polym Phys Ed 26:2093
45. Iliopoulos I, Audebert R (1991) Macromolecules 24:2566
46. Wang Y, Morawetz H (1989) Macromolecules 22:164
47. Yang T P, Pearce E, Kwei TK, Yang N (1989) Macromolecules 22:1813
48. Wang L, Pearce E, Kwei TK (1991) J Polym Sci B Polym Phys 29:619
49. Drago R, O'Bryan N, Vogel G (1970) J Am Chem Soc 92:3924
50. Drago R, Epley T (1969) J Am Chem Soc 91:2883
51. Suzuki T, Pearce E, Kwei TK (1992) Polymer 33:198
52. Joesten M, Schaad L (1974) In hydrogen bonding. Marcel Dekker, New York, p332
53. Sun J, Cabasso I (1989) J Polym Sci Polym Chem Ed 27:3985
54. Zhuang H, Pearce E, Kwei TK (1994) Macromolecules 27:6398
55. de Meftahi M, Fréchet M (1988) Polymer 29:477
56. Franzen M, Elliot J, Kyu T (1985) Macromolecules 28:5147
57. Lee J, Painter P, Coleman M (1988) Macromolecular 21:945
58. Dai J, Goh S, Lee S, Siow K (1994) Polymer J 26:905
59. Luo X, Goh S, Lee S (1997) Macromolecules 30:4934
60. Lin P, Clash C, Pearce E, Kwei TK, Aponte M (1988) J Polym Sci Polym Phys Ed 26:603
61. Kwei TK (1984) J Polym Sci Polym Lett Ed 22:307
62. Chatterjee S, Sethi K (1984) Polymer 25:1367
63. Dai J, Goh S, Lee S, Siow K (1994) Polymer 35:2174
64. Dai J, Goh S, Lee S, Siow K (1993) Polymer 34:4314
65. Cowie JMG (1979) Pure Appl Chem 51:2331
66. Bimendina L, Bekturov E, Tleubaeva G, Frolova V (1979) J Polym Sci Polym Symp 66:9
67. Pèrez-Dorado A, Pièrola I, Baselga J, Radic D (1989) Makromol Chem 190:2975
68. Pèrez-Dorado A, Pièrola I, Baselga J, Gargallo L (1990) Makromol Chem 191:2905
69. Meaurio E, Velade J, Cesteros L, Katime I (1996) Macromolecules 29:4598
70. Cesteros L, Velade J, Katime I (1995) Polymer 36:3183
71. Anasagasti M, Valenclano R, Bivas L, Katime I (1992) Polymer Bull 28:669
72. Cesteros L, Meaurio E, Katime I (1994) Polym Interl 34:97
73. Cesteros L, Rego J, Vazquez J, Katime I (1990) Polym Commun 31:152
74. Velada J, Cesteros L, Madoz A, Katime I (1995) Macromol Chem Phys 196:3171
75. Walsh D, Rostami S (1985) Adv Polym Sci 70:119
76. Olabisi O, Robeson L, Shaw M (1977) Polymer-polymer miscibility, Academic Press, New York

77. Eisenberg A, Smith P, Zhow ZL (1982) Polym Eng Sci 22:455
78. Zhou Z, Eisenberg A, (1983) J Polym Sci Polym Phys Ed 21:595
79. Smith P, Hara M, Eisenberg A (1987) In: Ottenbrite R, Utracki L, Inoue T (eds) Current topics in polymer science. vol 2. p 265
80. Ting S, Pearce E, Kwei TK (1980) J Polym Sci Polym Lett Ed 18 : 201
81. Ting S, Bulkin B, Pearce E, Kwei TK (1981) J Polym Sci Polym Chem Ed 19:1451
82. Coleman M, Graf J, Painter P (1991) Specific interactions and the miscibility of polymer blends. Technomic, Lancaster, PA
83. Coleman M, Painter P (1995) Prog Polym Sci 20:1
84. Yang X, Painter P, Coleman M, Pearce E, Kwei TK (1992) Macromolecules 25:2156
85. Pearce E, Kwei TK, Min B (1984) J Macromol Sci Chem A21:1181
86. Chu E, Pearce E, Kwei TK, Yeh T, Okamoto Y (1991) Macromol Chem Rapid Commun 12:1
87. Pearce E, Kwei TK (1992) In: Noda I, Rubingh D (eds) Polymer solutions, blends and interfaces. Elsevier, Amsterdam, pp 133–149
88. Purcell K, Stickeleather J, Brunk S (1969) J Am Chem Soc 91:4019
89. Jong L, Pearce E, Kwei TK, (1993) Polymer 34:48
90. Luo D, Pearce E, Kwei TK (1993) Macromolecules 26:6220
91. Cao X, Jiang M, Yu T (1989) Makromol Chem 190:117
92. Jiang M, Cao X, Chen W, Xiao H, Jin X, Yu T (1990) Makromol Chem Macromol Symp 38:161
93. Jiang M (1991) Chem J Chinese Univ 12:127
94. Odian G (1981) Principles of Polymerization, 2nd ed. Wiley, New York, chap 2
95. Kwei TK, Pearce E, Ren F, Chen J (1986) J Polym Sci Part B Polym Phys 24:1597
96. Cowie J, Reilly A (1993) J Appl Polym Sci 47:1155
97. Cowie J, Reilly A (1992) Polymer 33:4814
98. Cowie J, Reilly A (1993) Eur Polym J 29:455
99. Vermeesch I, Groeninckx G (1995) Polymer 36:1039
100. Lu S, Pearce E, Kwei TK (1994) J Macromol Sci Pure Appl Chem A31:1535
101. Lu S, Pearce E, Kwei TK (1993) Macromolecules 26:3514
102. Lu S, Pearce E, Kwei TK (1994) J Polym Sci Polym Chem Ed 32:2607
103. Lu S, Pearce E, Kwei TK (1994) Polym Eng Sci 35:1113
104. Trent J, Scheinbeim J, Couchman P (1983) Macromolecules 16:589
105. Morawetz H (1983) Polym Eng Sci 23:689
106. Amrani F, Hung J, Morawetz H (1980) Macromolecules 13:649
107. Chen C-T, Morawetz H (1989) Macromolecules 22:159
108. Jiang M, Chen W, Yu T (1991) Polymer 32:984
109. Schent W, Reichet D, Schneider H (1990) Polymer 31:29
110. Jong L, Pearce E, Kwie TK, Dickinson L (1990) Macromolecules 23:5071
111. Jiang M, Qiu X, Qin W, Fei L (1995) Macromolecules 28:730
112. Campbell G, VanderHart D, Feng Y, Han CC (1992) Macromolecules 25:2107
113. Taylor-Smith R, Register R (1993) Macromolecules 26:2802
114. Register R, Bell T (1992) J Polym Sci Part B:Polym Phys 30:569
115. Sperling L (1981) Interpenetrating polymer networks and related materials. Plenum, New York
116. Sperling L (1989) In: Allen G, Bevington J (eds) Comprehensive polymer science. Pergamon, Oxford, 6:423
117. Nishi S, Kotaka T(1985) Macromolecules 18: 1519
118. Bauer B, Briber R, Han C (1989) Macromolecules 22:940
119. Coleman M, Serman C, Painter P (1987) Macromolecules 22:226
120. Kim H, Pearce E, Kwei TK(1989) Macromolecules 22:3374
121. Jiang M, Xiao H, Jin X, Yu T (1990) Polym Bull 23:103
122. Jiang M, Xiao H, Yu T (1992) Polym Bull 29:434
123. Xiao H, Jiang M, Yu T (1994) Polym 25:5523

124. Xiao H, Jiang M, Yu T (1994) Polym 25:5529
125. Xiao H, Jiang M, Yu T (1993) Chem J Chinese Univ 14:1167
126. Jiang M, Xie H (1991) Prog Polym Sci 16: 977
127. Tucker P, Barlow J, Paul D (1988) Macromolecules 21:1678
128. Tucker P, Barlow J, Paul D (1988) Macromolecules 21:2794
129. Feng H, Feng Z, Ruan H, Shen L (1992) Macromolecules 25:5981
130. de Araujo M, Studler R, Cantow H (1988) Polymer 29:2235
131. Mitchell G, Windle A (1985) J Polym Sci Polym Phys Ed 23:1967
132. Xie H, Liu Y, Jiang M, Yu T (1986) Polymer 27:1928
133. Xie H, Liu Y, Jiang M, Yu T (1988) Makromol Chem Rapid Commun 9:79
134. Xie H, Liu Y, Jiang M, Yu T (1989) Makromol Chem Rapid Commun 10:115
135. Jiang M, Huang T, Xie J (1995) Macromol Chem Phys 196:787
136. Jiang M, Huang T, Xie J (1995) Macromol Chem Phys 196:803
137. Bosma M, ten Brink G, Elis T (1988) Macromolecules 21:1465
138. Grooten R, ten Brink G (1989) Macromolecules 22:1761
139. Huang T, Xie J, Jiang M (1995) Chem J Chinese Univ 16:302
140. Zhu Z, Zhang Y, Xie J, Jiang M (1997) Chem J Chinese Univ 18:1378
141. Jiang M (1997) Macromol Symp 118:377
142. Jiang M, Li M, Liu L, Xiang M, Zhu L (1997) Macromol Symp 124:135
143. Qiu X, Jiang M (1994) Polymer 35:5084
144. Jiang M, Qiu X (1993) Proceedings of the 34th IUPAC Congress Beijing, p 643
145. Qiu X, Jiang M (1995) Polymer 36:3601
146. Xiang M, Jiang M, Zhang Y, Wu C, Feng L (1997) Macromolecules 30:2313
147. Xiang M, Jiang M, Zhang Y, Wu C (1997) Macromolecules 30:5339
148. Zhu L, Jiang M, Liu L, Zhou H, Fan L, Zhang Y, Zhang YB, Wu C J (1998) Macromol Sci Phys B37(6):805
149. Zhu L, Jiang M, Liu L, Zhou H, Fan L, Zhang Y J (1998) Macromol Sci Phys B37(6):827
150. Zhou H, Xiang M, Chen W, Jiang M (1997) Macromol Chem Phys 198:809
151. Ringsdrof H, Simon J, Winnik F (1992) Macromolecules 25:5353
152. Zhang Y, Xiang M, Jiang M, Wu C (1997) Macromolecules 30:2035
153. Zhang Y, Xiang M, Jiang M, Wu C (1997) Macromolecules 30:6084
154. Zhou H (1997) PhD Thesis, Fudan University, Shanghai, China
155. Nagata M, Fukuda T, Inagaki (1987) Macromolecules 20:2173
156. Sun Z, Wang CH (1996) Macromolecules 29:2011
157. Xiang M, Jiang M, Feng L (1995) Macromol Rapid Commun 16:477
158. Zhu L, Liu L, Jiang M (1996) Macromol Rapid Commun 17:37
159. Zhu L (1996) M.S. Thesis, Fudan University, Shanghai, China
160. Liu L, Jiang M (1995) Macromolecules 28:8702
161. Perrin D, Dempsey B, Serjeant E (1981) PKa prediction for organic acids and bases. Chapman & Hall, New York
162. Djadoun S, Goldberg R (1977) Macromolecules 10:1015
163. Landry C, Tecgarden D (1991) Macromolecules 24:4310
164. Otocka E, Eirich F (1968) J Polym Sci Part A-2 6:895
165. Otocka E, Eirich F (1968) J Polym Sci Part A-2 6:913
166. Lindemann R, Zundel G (1972) J Chem Soc, Faraday Trans 2, 78:979
167. Smith P, Eisenberg A (1994) Macromolecules 27:545
168. Cook D (1961) Can J Chem 39:2009
169. Qiu X, Jiang M (1993) Chem J Chinese Univ 14:1625
170. Kwei TK, Pearce E, Pennacchia J, Charton M (1987) Macromolecules 20:1174
171. Goldstein M (1985) Macromolecules 18:277
172. Painter P, Graf J, Coleman M (1991) Macromolecules 24: 5630
173. Couchman P (1984) Polym Eng Sci 24:135
174. Martuscelli E, Scllitti C, Silvestre C (1985) Makromol Chem Rapid Commun 6:125
175. Zupper M, Simon G, Cherry P, Hill A (1994) J Polym Sci Part B Polym Phys 31:1237

176. Li W, Jin X, Jiang B (1989) Acta Polym Sin 337
177. Li W, Jin X, Jiang B (1990) Acta Polym Sin 336
178. Wang Z, Jiang B (1997) Macromolecules 30:6223
179. Zhang R, Luo X, Ma D (1995) J Appl Polym Sci 55:455
180. Ma D, Luo X, Zhang R, Nishi T (1996) Polym 37:1575
181. Vaidya M, Levon K, Pearce E (1995) J. Polym. Sci. Part B Polym. Phys. 33:2093
182. de Juana R, Cortázar M (1993) Macromolecules 26:1170
183. Fernández-Berridi M, Valero M, Martínnez de Iiardruya A, Espí E, Iruin J, (1993) Polymer 34:38
184. Khambatta F, Warner F, Russell T, Stein R (1976) J Polym Sci Polym Phys Ed 14:1391

Editor: Prof. H. Fujita
Received: July 1998

Author Index Volumes 101–146

Author Index Volumes 1–100 see Volume 100

Grosberg, A. and *Nechaev, S.*: Polymer Topology. Vol. 106, pp. 1-30.
Grubbs, R., Risse, W. and *Novac, B.*: The Development of Well-defined Catalysts for Ring-Opening Olefin Metathesis. Vol. 102, pp. 47-72.
van Gunsteren, W. F. see Gusev, A. A.: Vol. 116, pp. 207-248.
Gusev, A. A., Müller-Plathe, F., van Gunsteren, W. F. and *Suter, U. W.*: Dynamics of Small Molecules in Bulk Polymers. Vol. 116, pp. 207-248.
Guillot, J. see Hunkeler, D.: Vol. 112, pp. 115-134.
Guyot, A. and *Tauer, K.*: Reactive Surfactants in Emulsion Polymerization. Vol. 111, pp. 43-66.

Hadjichristidis, N., Pispas, S., Pitsikalis, M., Iatrou, H., Vlahos, C.: Asymmetric Star Polymers Synthesis and Properties. Vol. 142, pp. 71-128.
Hadjichristidis, N. see Xu, Z.: Vol. 120, pp. 1-50.
Hadjichristidis, N. see Pitsikalis, M.: Vol. 135, pp. 1-138.
Hall, H. K. see *Penelle, J.*: Vol. 102, pp. 73-104.
Hammouda, B.: SANS from Homogeneous Polymer Mixtures: A Unified Overview. Vol. 106, pp. 87-134.
Harada, A.: Design and Construction of Supramolecular Architectures Consisting of Cyclodextrins and Polymers. Vol. 133, pp. 141-192.
Haralson, M. A. see Prokop, A.: Vol. 136, pp. 1-52.
Hawker, C. J. see Hedrick, J. L.: Vol. 141, pp. 1-44.
Hedrick, J. L., Carter, K. R., Labadie, J. W., Miller, R. D., Volksen, W., Hawker, C. J., Yoon, D. Y., Russell, T. P., McGrath, J. E., Briber, R. M.: Nanoporous Polyimides. Vol. 141, pp. 1-44.
Hedrick, J. L. see Hergenrother, P. M.: Vol. 117, pp. 67-110.
Hedrick, J.L. see McGrath, J. E.: Vol. 140, pp. 61-106.
Heller, J.: Poly (Ortho Esters). Vol. 107, pp. 41-92.
Hemielec, A. A. see Hunkeler, D.: Vol. 112, pp. 115-134.
Hergenrother, P. M., Connell, J. W., Labadie, J. W. and *Hedrick, J. L.*: Poly(arylene ether)s Containing Heterocyclic Units. Vol. 117, pp. 67-110.
Hernández-Barajas, J. see Wandrey, C.: Vol. 145, pp. 123-182.

Hervet, H. see Léger, L.: Vol. 138, pp. 185-226.
Hiramatsu, N. see Matsushige, M.: Vol. 125, pp. 147-186.
Hirasa, O. see Suzuki, M.: Vol. 110, pp. 241-262.
Hirotsu, S.: Coexistence of Phases and the Nature of First-Order Transition in Poly-N-isopropylacrylamide Gels. Vol. 110, pp. 1-26.
Hornsby, P.: Rheology, Compoundind and Processing of Filled Thermoplastics. Vol. 139, pp. 155-216.
Hult, A., Johansson, M., Malmström, E.: Hyperbranched Polymers. Vol. 143, pp. 1-34.
Hunkeler, D., Candau, F., Pichot, C., Hemielec, A. E., Xie, T. Y., Barton, J., Vaskova, V., Guillot, J., Dimonie, M. V., Reichert, K. H.: Heterophase Polymerization: A Physical and Kinetic Comparision and Categorization. Vol. 112, pp. 115-134.
Hunkeler, D. see Prokop, A.: Vol. 136, pp. 1-52; 53-74.
Hunkeler, D see Wandrey, C.: Vol. 145, pp. 123-182.

Iatrou, H. see *Hadjichristidis, N.*: Vol. 142, pp. 71-128.
Ichikawa, T. see Yoshida, H.: Vol. 105, pp. 3-36.
Ihara, E. see Yasuda, H.: Vol. 133, pp. 53-102.
Ikada, Y. see Uyama, Y.: Vol. 137, pp. 1-40.
Ilavsky, M.: Effect on Phase Transition on Swelling and Mechanical Behavior of Synthetic Hydrogels. Vol. 109, pp. 173-206.
Imai, Y.: Rapid Synthesis of Polyimides from Nylon-Salt Monomers. Vol. 140, pp. 1-23.
Inomata, H. see Saito, S.: Vol. 106, pp. 207-232.
Inoue, S. see Sugimoto, H.: Vol. 146, pp. 39-120.
Irie, M.: Stimuli-Responsive Poly(N-isopropylacrylamide), Photo- and Chemical-Induced Phase Transitions. Vol. 110, pp. 49-66.
Ise, N. see Matsuoka, H.: Vol. 114, pp. 187-232.

Subject Index